Betty B. Schechter, Publisher
Barbara K. Baker, Marketing Manager
Mark Linton, Developmental Editor
Martha Conway, Production Editor
Karen Roberts, Production Coordinator

I(T)P
International Thomson Publishing Company
South-Western Educational Publishing is an ITP company.
The ITP Trademark is used under license.

ISBN: 0 - 538 - 63995 - 4
1 2 3 4 5 6 7 8 BN 99 98 97 96 95
Printed in the United States of America

This book is printed on recycled, acid-free paper that meets Environmental Protection Agency standards.

South-Western Educational Publishing

Pre-GED Science Exercises

INTRODUCTION

Welcome to South-Western's *Pre-GED Science Exercises* book—an important part of the South-Western GED Preparation System. South-Western's GED Preparation System includes test preparation books, exercises books, and state-of-the-art interactive software. Each of these learning tools is available at two levels—Pre-GED and GED.

WHAT WILL I FIND IN THIS BOOK?

This *Pre-GED Science Exercises* book is really an extension of the South-Western *Pre-GED Science* book. It offers additional practice in the subject areas covered on the GED Science Test. The exercises reinforce the thinking skills you learned in Section 2 of the *Pre-GED Science* book. This book also contains optional writing activities and a posttest.

EXERCISES

The four units of exercises in this book correspond to the units found in Section 3 of the South-Western *Pre-GED Science* book.

OPTIONAL WRITING ACTIVITIES

Each unit ends with one or more optional writing activities. These writing exercises will give you opportunities to write about science topics as they relate to your daily life.

POSTTEST

A posttest is included at the back of this book. It provides an additional opportunity to increase your test-taking skills. It also helps you evaluate your progress toward your goal—success on the GED Science Test!

Part I contains questions in the same format as many of the exercises in this book—fill-in-the-blanks, matching, true and false. Part II contains questions in the same format as the GED—multiple choice. Part II is about one-half the length of the actual GED Science Test. You can take the two parts together or at different times.

ANSWERS AND EXPLANATIONS

The answers and explanations for the units are provided near the back of the book. They will help you better understand each question. Check your answers after completing each exercise. The posttest answers and explanations come after the posttest. Use the Referral Charts at the end of the posttest answers and explanations to check your success.

HOW CAN I USE THIS BOOK?

You can use this book in one of two ways. The most effective way is to use it with the South-Western *Pre-GED Science* book. Complete the individual exercises as you work through each Section 3 science lesson. Another way to use this book is to complete the exercises as a review or checkup on your skill development. Either way, you should take the posttest. Then, complete the Referral Charts at the end of the posttest answers and explanations. Study the results carefully to see your strengths. Use the results to determine where you may need additional practice before moving on to GED-level materials.

Work with your instructor to determine if you are ready to move on to the South-Western *GED Science* book. This will be your next step in preparing for the GED Tests. The evaluation charts at the end of this book will help you track your progress.

WHERE DO I GO FOR MORE HELP?

For additional instruction and practice, you can use South-Western's Pre-GED *Advantage* software. It is one more way to give you the *connection to success* as you prepare for the GED Tests!

ACKNOWLEDGEMENTS

Art:

Network Graphics: 5, 6, 8, 9, 10, 12, 19, 25, 30, 35, 40, 43, 51, 56, 57, 59, 60, 79, 81

Carol Porteous: 18, 20, 26, 27, 31, 33, 45, 49, 74, 76, 80

The Structure of Cells: Organelles

Science passages may contain a lot of new vocabulary. To find the meaning of new words, you have to look for details. As you read the passage below, use the question "what" to help you with the meaning of the italicized words. For example, ask yourself, "What is a nucleus?" "What is its job in the cell?"

Cells are the smallest units, or parts, of living things. They perform all of the functions necessary for life. Each cell is made up of parts which perform different jobs. Together, these parts allow the cell to stay alive.

The different parts of a cell are called *organelles*, or "tiny organs." One organelle, the *nucleus*, is responsible for cell reproduction. The nucleus holds the "blueprints," or plans, for all the chemicals the cell produces. Because of this, we can say it directs all of the cell's activities. Other organelles, the *ribosomes,* are the cell's "factories." They use some of the cell's energy to make proteins. Still other organelles, called *mitochondria*, are the powerhouses of the cells. They release energy from food.

■ *For **Items 1 to 4**, match each term with its correct definition.*

	Term		Definition
_____ 1.	organelle	a.	cell's director, holds the blueprints
_____ 2.	nucleus	b.	cell's powerhouses, release energy from food
_____ 3.	ribosome	c.	smallest part or unit of any living thing
_____ 4.	mitochondria	d.	cell's factories, use energy to make proteins

■ *Use the information in the passage to answer **Items 5 and 6**.*

5. Each *organ* of the body performs a certain job or jobs. This allows the body to live. For example, the liver stores a chemical called glycogen. Why, then, is it correct to call the small parts of the cell *organelles*?

____

____

6. Skin cells produce more skin cells. And muscle cells produce more muscle cells. Which organelle is responsible for these processes? Explain.

____

____

► *For more information about this topic, see pages 138–139 in Pre-GED Science. Answers begin on page 65.*

Plant Cells: Organelles

When you *compare* two things, you state how they are alike. When you *contrast* them, you state how they are different. In the exercise below, you will compare and contrast plant and animal cells.

In general, plant and animal cells perform the same functions. Because of this, they have many structures, or organelles, in common. These include the nucleus, cell membrane, ribosomes, and mitochondria. However, plant and animal cells differ in some important ways.

For example, plant cells have rigid *cell walls*. Animal cells have no cell walls. In plants, the cell wall surrounds the cell membrane. It provides support and protection for the cell. Plant cells also have organelles called *chloroplasts*. The chloroplasts contain molecules of the substance chlorophyll. This organelle allows the plant to use the sun's energy to make its own food.

■ *Use the information above to answer **Items 1 to 5**.*

1. Name two organelles found in plants but not in animals.

____

2. Name two organelles found in both plants and animals.

____

3. What are two functions of the cell wall?

____

4. What is the difference between chloroplasts and chlorophyll?

____

► *For more information about this topic, see pages 140–141 in Pre-GED Science. Answers begin on page 65.*

Specialized Cells in Humans: Cancer

When you *restate* facts, you repeat information using your own words. As you answer the questions below, try to use your own words instead of just copying the information in the passage.

In addition to red blood cells, nerve cells, and muscle cells, your body contains a wide variety of cells that do different and special jobs. *Cancer* is a disease in which cells lose their ability to perform their special jobs. But, like all cells, cancer cells grow and multiply. What's more, cancer cells generally multiply faster than normal cells. As the cancer cells grow, they crowd out the normal cells. If this happens in your lungs, for instance, your ability to breathe will be threatened.

What makes normal cells turn into cancer cells? Medical scientists do not know for certain. There are many possible factors, or influences. For example, chemicals found in tobacco can cause lung cancer. Overexposure to sunlight can cause skin cancer. Some viruses can cause cancers in other parts of the body. Today, scientists believe that most cancers are caused by a combination of several factors working together.

■ *Use the information above to answer **Items 1 to 5**.*

1. Name three kinds of specialized cells.

____

2. In what ways are cancer cells different from specialized cells?

____

3. In what ways are cancer cells similar to specialized cells?

____

4. When lung cells become cancerous, what special job of the lung is endangered?

____

5. What is one factor that might cause lung cancer? Skin cancer?

____

► *For more information about this topic, see page 142 in Pre-GED Science. Answers begin on page 65.*

Specialized Cells in Humans: AIDS and T-4 Lymphocytes

The *main idea* of a passage is the most important idea or ideas found in it. The following passage has two main ideas: how HIV causes illness and how AIDS is spread. As you read, watch for details that support these ideas.

White blood cells are part of your immune system. This system protects your body by attacking microorganisms that cause diseases. However, a microorganism called the *human immunodeficiency virus* (HIV) destroys a very important part of the immune system. This part is a white blood cell called a *T-4 lymphocyte* (sometimes called a helper T-cell). T-4 lymphocytes produce antibodies that fight infections.

As time passes, HIV destroys more and more T-4 lymphocytes. The body's immune system breaks down. That is, its ability to fight off diseases becomes weaker and weaker. Microorganisms that would ordinarily do little damage can now cause life-threatening infections. At this point, a person is said to have AIDS (Acquired Immune Deficiency Syndrome). He or she has acquired, or caught, a disease that weakens the ability to resist infections.

HIV can be transmitted through unprotected sexual contact, blood transfusions, and the sharing of needles by infected persons. The virus also can be passed from a mother to her unborn child or through a mother's milk to her nursing child.

■ *Use the information above to answer* ***Items 1 to 4.***

1. Which family of blood cells defends the body against infections?

____

2. What kind of microorganism causes AIDS?

____

3. What specialized job do T-4 lymphocytes do in the body?

____

4. Explain how HIV affects the immune system.

____

► *For more information about this topic, see page 142 in Pre-GED Science.*
Answers begin on page 65.

Unit 1 **LESSON 2 • EXERCISE 1**

The Respiratory System: Breathing Rates

When you *infer,* you find out something that is not actually stated in words or shown visually. As you read the passage and study the graph below, you use what you know to understand something hinted at but not said.

To carry out all the activities of a living organism, your body requires a gas, oxygen. All body cells need oxygen to produce energy. As physical activity increases, the body needs more energy. Since the cells have to produce more energy, more oxygen is needed. Your breathing rate will increase so you can take in more oxygen. The graph shows changes in a person's breathing rate over a period of 24 minutes.

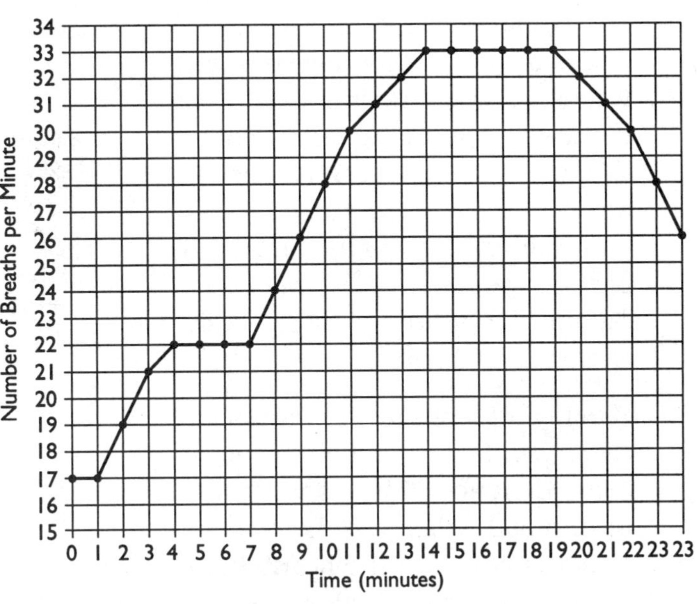

■ *Use the graph and information above to answer* ***Items 1 to 4.***

1. What does the information down the left side of the graph represent? What does the information across the bottom represent?

____

2. What is the lowest number of breaths per minute entered on the graph?

____

3. How many breaths per minute were taken at the 8-minute mark?

____

4. Do you think the person was probably sitting or walking between minutes 7-13? Explain your answer.

____

____

► *For more information on the Respiratory System see pages 144–145 in* *Pre-GED Science.* *Answers begin on page 65.*

The Circulatory System: The Heart

Some diagrams show a process or series of steps. To understand this type of diagram, first read any labels. Then look for information, such as numbers or arrows, that shows the sequence or the order in which events take place.

The *circulatory system* is made up of the heart, the blood vessels, and the blood. Your heart muscle contracts to push the blood along through the blood vessels. The heart is actually two pumps. One part pumps blood to the lungs and back to the heart. The other one pumps blood throughout the body and back to the heart.

When your heart muscle contracts, each chamber, or section of the heart, contracts. The contractions push blood either into the next chamber or out of the heart. Blood can flow in one direction only. This is because the heart is equipped with one-way valves. The arrows in the drawing show the direction blood flows through the heart. The drawing also shows the location of the heart valves.

INTERIOR OF HUMAN HEART—FRONT VIEW

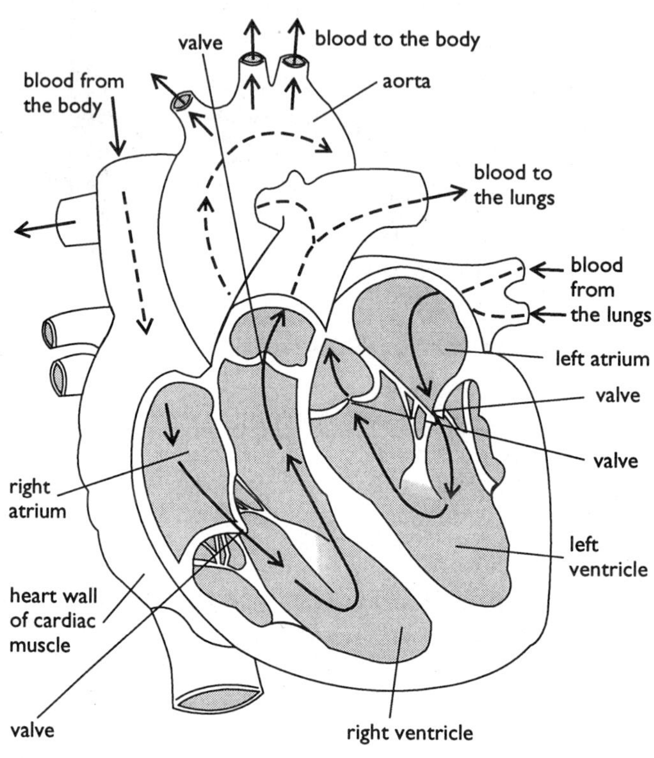

■ *Use the information and drawing above to complete* ***Items 1 to 6.***

1. The four chambers of the heart are the ______________, ______________, ______________, and ______________.
2. Blood from the body enters the ______________ of the heart.
3. When the right ventricle contracts, the blood flows to ______________________.
4. The purpose of the valves is to ______________________________.
5. Blood flows from the left atrium to the ______________________.
6. The heart's job is to ______________________________.

▶ *For more information about this topic see page 146 in Pre-GED Science. Answers begin on page 65.*

The Circulatory System: Blood Pressure

A *fact* is a piece of information that is true. The facts needed to answer the questions below are clearly stated in the passage. Before you answer a question, go back to the passage and find the fact related to the question.

As your heart contracts, or beats, blood is pushed throughout your body. The measurement of this pushing force is called your *blood pressure*. Blood pressure is given in two numbers. The first number indicates the pressure when your heart contracts. The second number gives the pressure when your heart is relaxed, or between beats. Since the push is stronger when your heart contracts, the first number is always higher than the second number.

These numbers are important. They tell whether a person's blood pressure is normal, above normal, or below normal. People who have very high blood pressure can have serious health problems. Strong pressure can break blood vessels that lead to the brain, causing a stroke. Strokes can cause brain damage and sometimes death. People who always have low blood pressure may lead normal, healthy lives. But if the pressure drops suddenly, it can be a sign of bleeding somewhere inside the body. People who have constant low blood pressure may feel tired or have frequent fainting spells.

Normal blood pressure for adults is about 120/80. If the first number is higher than 150, doctors consider it a health risk. If the second figure is greater than 90, the cause should be investigated.

■ *For **Items 1 to 4**, write* T *in the space provided if the statement is true. Write* F *if it is false. If the statement is false, change the underlined word to make the statement true. Write the new word in the space provided.*

_____ 1. The <u>higher</u> number given for blood pressure corresponds to the pressure when the heart is contracting. ______________________

_____ 2. A blood pressure of <u>90/55</u> is considered normal. ______________________

_____ 3. A sudden <u>rise</u> in blood pressure may indicate internal bleeding or bleeding inside the body. ______________________

_____ 4. <u>Low</u> blood pressure could lead to brain damage. ______________________

► *For more information about this topic, see page 146 in* <u>*Pre-GED Science*</u>. *Answers begin on page 65.*

The Skeletal System: Joints

Sometimes you need information from both a passage and an illustration. Read the passage and then look at the illustration to see how they are related. If an example is given, be certain you understand the example.

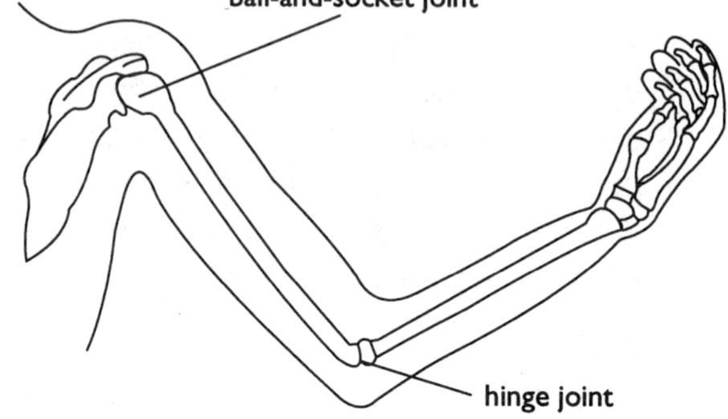

If all the bones in your body were cemented together, you would not be able to move. However, most of your bones come together at movable *joints*, such as your shoulder, elbow, hip, and knee. Strong ligaments hold the bones together at these joints.

There are four kinds of movable joints in your body. In *ball-and-socket* joints, the rounded end of one bone fits into a hollow in the adjoining bone. Many radios have ball-and-socket joints where the antenna is attached. These joints allow a wide range of movement. A *hinge* joint is like a hinge on a door. It permits movement. *Pivot* joints allow a rotating movement from side-to-side or up-and-down. In a *gliding* joint, almost flat-surfaced bones slide past one another. Gliding joints are found in the wrists and ankles.

■ *Use the information and illustration above to answer **Items 1 to 5**.*

1. Which joint, the shoulder or the elbow, is a ball-and-socket joint? ____________________
2. Which joint, the hip or the knee, is a hinge joint? ____________________
3. Which kind of joint permits large circular movements? ____________________
4. Which kind of joint allows movement back and forth? ____________________
5. What kind of joint do you think connects your skull to your spine? ____________________

▶ *For more information about this topic, see page 147–148 in Pre-GED Science. Answers begin on page 66.*

Unit 1 LESSON 2 • EXERCISE 5

The Skeletal System: Muscles Pull on Bones

A *cause* is something that makes something else happen. An *effect* is what happens. Use the information and illustration below to understand cause-and-effect relationships between muscles and bones.

Muscles that move bones are called *skeletal muscles*. They are attached to bones by tissue called *tendons*. When you walk, wave, or point a finger, a muscle (or muscles) *contracts*, or bunches up. These contractions pull on bone. This pull makes the bone move at a joint.

Skeletal muscles usually work in pairs. If one of the pair is contracted, the other is relaxed. The drawing shows a muscle pair of the upper arm, the biceps, and the triceps.

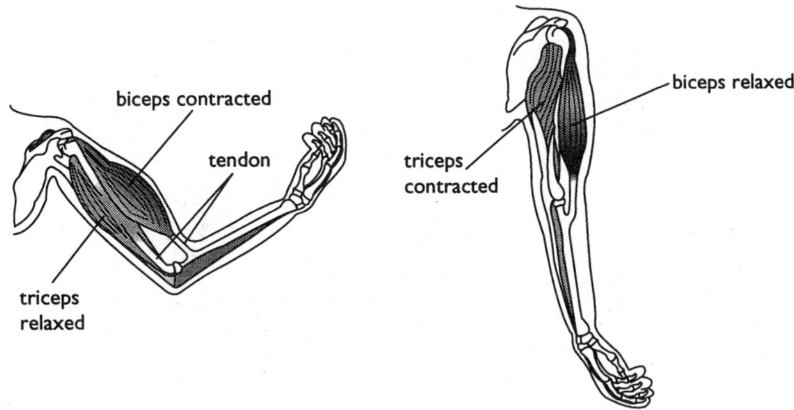

■ *Use the information and illustration above to answer* ***Items 1 to 4***.

1. When the arm is bent, is the biceps muscle contracted or relaxed?

____

2. How is the biceps muscle attached to the lower arm?

____

3. What happens to a muscle when it contracts?

____

4. How do muscle contractions cause bones to move?

____

► *For more information about this topic, see pages 147–148 in* Pre-GED Science.
Answers begin on page 66.

Reproductive Systems: Ovulation and Menstruation

A large amount of information can be shown in a diagram. To interpret a diagram, first read the title. Then read any labels on the diagram. Try to understand words that are new to you. Look for text that may help you understand the illustration.

Starting at puberty, hormonal changes in young women cause ovulation and menstruation to begin. Every 28 to 30 days, an egg is released from the ovaries. This release is called *ovulation*. If pregnancy does not occur, the unfertilized egg and the thickened lining of the uterus are shed through the vagina. This process is called *menstruation*.

The diagram shows changes in the thickness of the lining of the uterus during the menstrual cycle. It also shows when ovulation takes place.

MENSTRUAL CYCLE

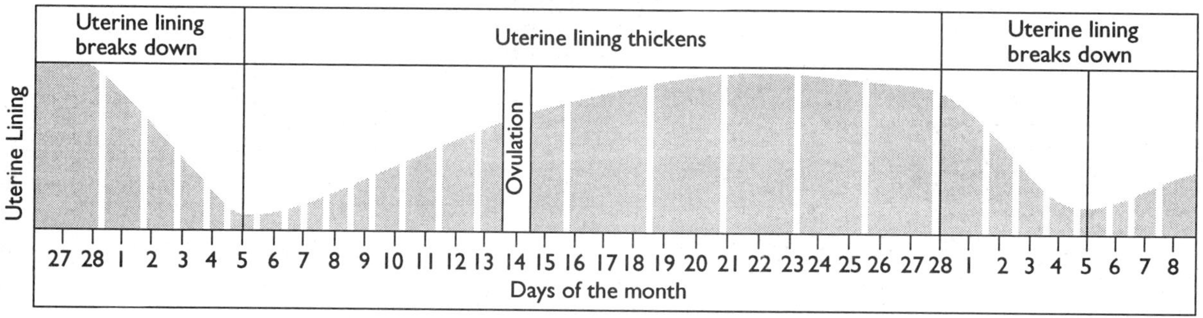

■ *Use the information above to answer **Items 1 to 5**.*

1. What is ovulation?

____

____

2. What is menstruation?

____

____

3. The diagram shows a menstrual cycle whose length is __________ days.

4. On about what day is the lining of the uterus thickest? __________

5. In this cycle, the egg is released on about what day? __________

► *For more information about this topic, see pages 150–151 in Pre-GED Science. Answers begin on page 66.*

Fetal Development: Growth in the Uterus

Supporting details in a passage help you understand more about the main idea. They tell more about the topic but do not introduce a new topic. Notice the supporting details for the main idea of the passage below.

Fertilization occurs when a sperm enters an egg. This usually occurs in a woman's Fallopian tube. During the next six days, the fertilized egg moves through the tube and into the uterus. As the egg moves, it divides many times, forming a hollow ball of cells. This ball of cells, or *embryo*, becomes attached to the lining of the uterus. During the next nine months, the embryo will develop into a *fetus*.

The fetus needs oxygen and food in order to grow and develop. It also must get rid of the wastes that it produces. How does the fetus take in oxygen and food and rid itself of wastes? These processes are made possible by a temporary organ called the *placenta*. Blood vessels from the mother enter the placenta on one side. Blood vessels from the fetus pass through an *umbilical cord* attached to the fetus. They enter the placenta from the other side.

The mother's and the fetus' blood vessels are not connected. They are both connected to the placenta. Oxygen and digested foods can pass from the mother's blood vessels into those of the fetus. Wastes from the fetus can pass from its blood vessels into those of its mother.

■ *For **Items 1 to 3**, write* T *in the space provided if the statement is true. Write* F *if it is false. If the statement is false, change the underlined word to make the statement true. Write the new word in the space provided.*

_____ 1. The embryo attaches to the lining of the uterus. ______________

_____ 2. The umbilical cord is connected to the mother. ______________

_____ 3. The blood vessels of the mother and fetus are connected to each other. ______________

■ *Use the information in the passage above to answer **Item 4**.*

4. What does a fetus need to grow and develop?

____

____

▶ *For more information about this topic, see pages 152–153 in Pre-GED Science. Answers begin on page 66.*

Unit 1 LESSON 3 • EXERCISE 3

Heredity: "Strong" and "Weak" Genes

Science readings are full of details. Use the questions *who? what? where? when? why?* and *how?* to find details in the passage below. Remember, details give more information about a main idea.

You have 23 pairs of chromosomes. One of each pair came from your mother. The other came from your father. The chromosomes and the genes on them are equally paired. This means that you have two genes for each trait.

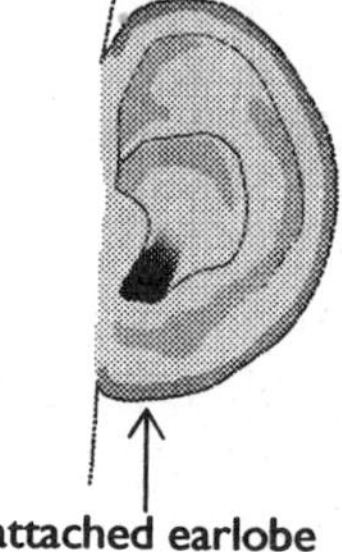

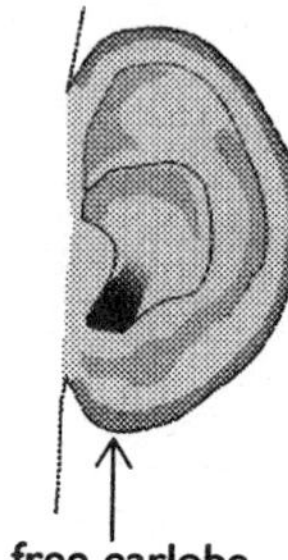

If you look at your earlobes in a mirror, you will see that they are either attached or free, as shown in the illustration. You received one gene for shape from your mother and the other gene from your father.

You can think of genes as having different strengths. The gene for free earlobes is "stronger," or *dominant*. The gene for attached earlobes is a "weaker" or *recessive* gene. Suppose you received one dominant and one recessive gene for earlobe shape. You would have free earlobes. This is because the free earlobe gene is a dominant gene. You would have attached earlobes only if you received a recessive attached earlobe gene from both of your parents.

■ *Use the information and illustration above to answer* ***Items 1 to 5***.

1. What is another name for a "strong" gene? ____________________

2. What is another name for a "weak" gene? ____________________

3. How many genes for earlobe shape did you inherit? ____________________

4. What is the shape of your earlobes? ____________________

5. For a person with free earlobes, what two combinations of genes (dominant and recessive) are possible?

____

► *For more information about this topic, see page 154 in Pre-GED Science. Answers begin on page 66.*

 LESSON 3 • EXERCISE 4

Heredity: Genetic Engineering

A *fact* is a piece of information that is proven to be true. An *opinion* is what a person thinks. Opinions may or may not be supported by facts. Keep this in mind as you work on this exercise.

The genes on chromosomes are made up of a chemical called *deoxyribonucleic acid,* or DNA. The structure of the DNA of each of the genes is different. It is this structure that determines what a gene does.

During the 1970s, scientists found a way to move pieces of DNA from one living thing to another. This technique is called *genetic engineering*. Genetic engineering has many practical and important uses. Scientists expect to find more uses.

You probably have heard of a condition called diabetes. The pancreas of a person who has diabetes is unable to make the chemical insulin. Insulin controls the concentration of sugar in a person's blood. Too much sugar in a person's blood can make a person very sick. So a lack of insulin is a serious medical problem.

For years, many diabetic people used insulin taken from cattle and pigs. Some people had allergic reactions to this insulin. The ideal kind of insulin to use was human insulin. Before genetic engineering, there was no practical way of getting or producing human insulin.

Using special methods, scientists discovered they could transfer a human gene to certain bacteria. (Bacteria are tiny organisms smaller than human cells.) By adding the human gene for making insulin to bacteria, scientists produced helpful bacteria. The DNA blueprint of the gene that had been put in the bacteria caused the bacteria to make human insulin. Today, many people who have diabetes give themselves injections of human insulin made by bacteria. There is plenty of the insulin to go around and very few people are allergic to it.

■ *For **Items 1 to 5**, write* F *if the statement is a fact or* O *if it is an opinion.*

_____ **1.** Transferring DNA from one organism to another is called genetic engineering.

_____ **2.** More uses will be found for genetic engineering.

_____ **3.** Diabetes is caused by the body's inability to produce insulin.

_____ **4.** The gene for making insulin can be added to bacteria.

_____ **5.** Genetically engineered bacteria might be dangerous.

► *For more information about this topic, see page 154 in Pre-GED Science.*
Answers begin on page 66.

Nutrition: Saturated Fats

Tables help you find information easily. Read the headings at the top of each column. Then use the table to compare and contrast the items in each row.

Evidence shows that a group of fats, called *saturated fats*, can cause diseases of the heart and blood vessels. If a person's diet is high in saturated fats, they can build up in the blood and stick to the walls of blood vessels. As time passes, the blood vessels become clogged. This can lead to many health problems.

The table shows the saturated fat content of a number of foods. Fruits and vegetables have little or no saturated fats and are not included.

SATURATED FAT CONTENT OF SOME FOODS

Food	Saturated Fat (%)	Food	Saturated Fat (%)	Food	Saturated Fat (%)
Meats		Dairy Products		Fish	
T-bone steak	18.0	cream cheese	21.0	herring	3.0
bacon	18.0	Swiss cheese	18.0	tuna, albacore (canned, light)	2.0
lamb (shoulder)	13.0	half-and-half cream	10.0	rainbow trout	1.0
pork loin	10.0	skim milk	0	red snapper	1.0
Cereals and Grains		Oils			
wheat germ	2.0	coconut	86.0		
oats (puffed)	1.0	olive	14.0		
rice	0.1	safflower	9.0		
corn	0	canola	7.0		

■ *Use the information above to answer* ***Items 1 to 4.***

1. Excluding oils, which two families of foods are richest in saturated fats?

____

2. Which family of foods is lowest in saturated fats?

____

3. Of the oils, which is the lowest in saturated fat?

____

4. How would substituting a serving of rainbow trout for an equal-sized serving of T-bone steak affect the amount of saturated fat you would eat?

____

► *For more information about this topic, see pages 156–157 in* Pre-GED Science. *Answers begin on page 66.*

LESSON 4 • EXERCISE 2

Nutrition: Minerals

Sometimes information is much easier to read when it is in table form. Whenever you need to read a table, look first at its title and column headings. Be certain you understand what type of information you will find in the table. Then compare and contrast the information listed in the rows.

The table below lists four minerals that are important to your health. It also lists foods that are rich in each mineral and what each mineral does in your body.

SOME MINERALS NEEDED FOR HEALTH

Mineral	Main Food Sources	Uses in Body
potassium	fruit, whole grains, meat	needed for steady heart beat, action of nerves, and maintaining water balance in body
calcium	dairy products, dark green vegetables	needed for bone and tooth building, action of nerves and muscles, blood clotting
magnesium	grains, nuts, seafood, dark green vegetables, dairy products	helps body store energy
iron	meats, dark green vegetables, dried fruits	needed to make red blood cells

■ *Use the table and the information above to answer* ***Items 1 to 4.***

1. Which family of foods is a good source of calcium, magnesium, and iron?

____

2. What mineral is needed to make healthy bones and teeth?

____

3. The making of healthy red blood cells requires which mineral?

____

4. Why is magnesium important to your body?

____

► *For more information about this topic, see pages 156–157 in Pre-GED Science. Answers begin on page 66.*

Unit 1 **LESSON 4 • EXERCISE 3**

Noncommunicable Diseases: Asthma

A *cause* is what makes something happen. An *effect* is the result of something that has happened. In medicine, understanding what causes an illness helps us find a cure for it.

Asthma is a noncommunicable condition. When a person with asthma inhales a substance to which he or she is allergic, the result is often an *asthma attack*. When this happens, a series of changes occurs in the person's breathing tubes. Muscles around the tubes may contract, making the tubes narrower. The lining of the tubes may become inflamed and swollen, also narrowing the tubes. More mucus may be produced inside the tubes. As the mucus builds up, the tubes are blocked even more.

In a mild attack, a person feels uncomfortable and coughs. In a more serious attack, there is difficulty breathing. A wheezing sound may be heard as air is pushed through clogged tubes. Very serious attacks require immediate care. Death will result if breathing tubes are totally blocked.

The following substances are common causes, or *triggers*, of asthma attacks: pollen from plants, dust or mold in the air, or tiny pieces of hair and skin from pets. The best treatment is to avoid substances that trigger attacks. This is not always possible. People with serious asthmatic conditions should keep medicines with them at all times. If the medicine does not work quickly, a to a hospital emergency room may be necessary.

■ *Use the information above to answer* ***Items 1 to 4.***

1. What part of the body is affected by asthma? ______________________

2. Review the first paragraph. What three things can happen during an asthma attack?

____

____

3. Look at the last paragraph. What are four substances that can cause asthma attacks?

____

4. Why should asthma be considered a serious health problem?

____

▶ *For more information about this topic, see pages 158–159 in Pre-GED Science.*
Answers begin on page 66.

 **LESSON 4 • EXERCISE 4**

Noncommunicable Diseases: Diseases of the Heart and Blood Vessels

A list is a type of graphic organizer. Often it is easier to find information in a list than in a paragraph. Active readers often take notes in list form. Notice how the list in the passsage below helps organize the information.

Almost all diseases of the heart and blood vessels are noncommunicable. One major condition is a heart attack. A heart attack causes loss of oxygen to the brain. This loss of oxygen may be caused by a broken or blocked blood vessel. A blockage can be caused by a blood clot or fatty deposits. Whatever the cause, it is important to know the warning signs of a heart attack. The following signs may occur by themselves or in various combinations. You should be especially concerned if discomfort in your chest is accompanied by one or more of the other signs.

WARNING SIGNS OF A HEART ATTACK

1. Pain or pressure in the center of the chest near the breast bone.
2. Pain in the neck, jaw, left shoulder or left arm.
3. Sudden heavy perspiration and a feeling of panic.
4. Nausea, vomiting along with one or more of the above warning signs.
5. Shortness of breath.
6. Dizziness, fainting.

■ *Use the information above to answer **Item 1**.*

1. What are two ways a loss of oxygen supply to the brain may occur?

____

■ *For **Items 2 to 4**, write* T *in the space provided if the statement is true. Write* F *if it is false. If the statement is false, change the underlined word to make the sentence true. Write the new word in the space provided.*

_____ 2. Pain or pressure in the center of the back is a warning sign of a heart attack.

_____ 3. Sudden heavy perspiration and a feeling of tiredness may be caused by a heart attack.

_____ 4. Pain in the left shoulder or arm may be signs of a heart attack.

► *For more information about this topic, see pages 158–159 in Pre-GED Science. Answers begin on page 66.*

Keeping Fit: A Fit Heart

Line graphs can be used to compare or contrast two or more pieces of information. Be certain to read the key so you understand what the graph represents.

When you are in good physical shape, your heart does not have to work as hard. If your heart does less work than you do, it is working more efficiently. If you huff and puff, it shows that your heart is working less efficiently. The goal of a good fitness program is to help your heart work more efficiently. The graph shows the heart rate of two people as they perform the same activities. Notice the difference in the heart rates.

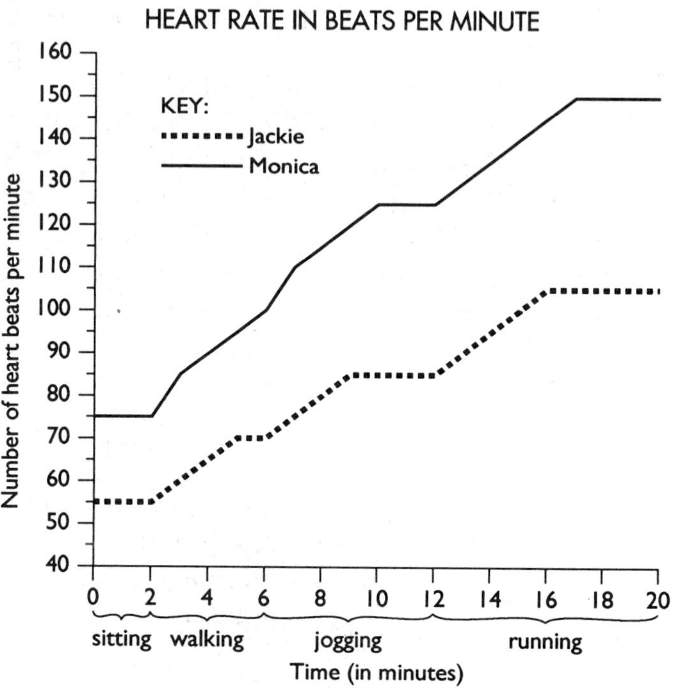

■ *Use the graph and information above to answer* ***Items 1 to 3.***

1. What was Jackie's heart rate when sitting? Monica's?

____

2. After how many minutes of jogging did Jackie's heart rate level off? Monica's?

____

3. What was Jackie's maximum heart rate? Monica's?

____

4. Whose heart is working more efficiently? Explain your answer.

____

____

▶ *For more information about this topic, see page 160 in* <u>*Pre-GED Science.*</u>
Answers begin on page 67.

Respiration: Breathing

A model is a device that shows how something works. It does not actually look like what it represents but is a simplified version of it. A model helps you understand a process. Look at the model below on the process of breathing.

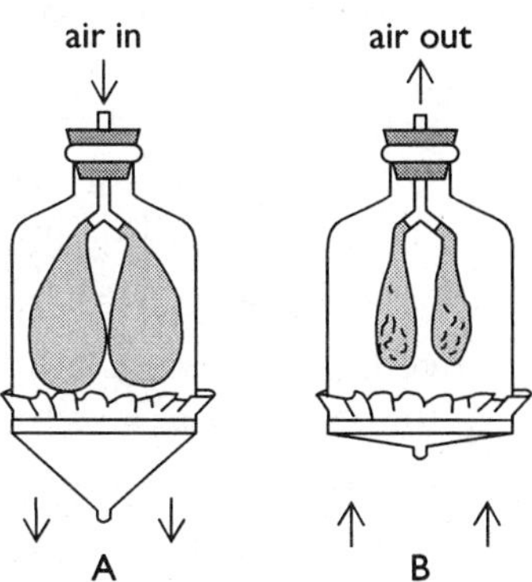

How, exactly, do you breathe? The illustration shows a model to help you understand the process.

The jar represents your chest. (Remember, however, that your chest can move in and out.) The piece of thin rubber at the open bottom of the jar represents a muscle called the *diaphragm*. The diaphragm separates your chest from your abdomen. Drawing A shows the position of the diaphragm when it is moved downward. The chest cavity becomes larger and the lungs fill with air. Drawing B shows the position of the diaphragm when it is relaxed. The chest cavity shrinks and the air is pushed from the lungs.

In addition to the movement of the diaphragm, the ribs move in and out. As the ribs expand outward, the chest cavity becomes larger. This also allows the lungs to expand and fill with air.

■ *For **Items 1 to 5**, circle the correct term in parentheses.*

1. When the diaphragm moves down, you breathe (in/out).
2. There is (more/less) space in a person's lungs when the diaphragm is relaxed.
3. The diaphragm is a (muscle/blood vessel).
4. As your inhale, the ribs move (inward/outward).
5. The balloons in the drawings represent the (ribs/lungs).

► *For more information about this topic, see pages 162–163 in Pre-GED Science. Answers begin on page 67.*

Photosynthesis: The Effect of Temperature on Photosynthesis

Line graphs are used to show information that changes over time or that changes as some condition changes. If the line moves up, it usually indicates an increase. Use the line graph below to understand photosynthesis.

Photosynthesis is the process in which plants use sunlight to produce food. Like all chemical reactions, the rate of photosynthesis is determined by a number of factors. One of these is temperature. The graph shows the change in rate of photosynthesis as the temperature changes.

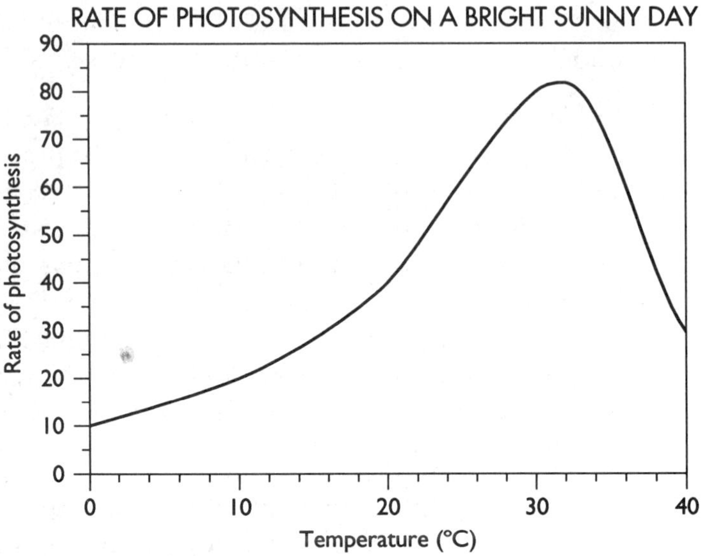

■ *Use the passage and the graph to answer* ***Items 1 to 3****. For* ***Items 1 and 2****, circle the correct answer. Use the answer blank for* ***Item 3****.*

1. At about what temperature is the rate of photosynthesis the highest?

   20°C 32°C 38°C

2. At about what temperature is the rate of photosynthesis the same as it is at 40°C?

   16°C 20°C 38°C

3. The graph shows rates of photosynthesis throughout a very bright, sunny day. Describe what the line on the graph would look like if the day were cloudy. Explain why you think this.

   ____

   ____

► *For more information about this topic, see pages 164–165 in* Pre-GED Science. *Answers begin on page 67.*

 **LESSON 5 • EXERCISE 3**

Photosynthesis: Comparing Photosynthesis and Respiration

The "5 W and H" questions—*who? what? where? when? why?* and *how?*—are used to find out details. Below, answers to some of these questions are organized in a chart.

Photosynthesis and respiration make life on Earth possible. Photosynthesis allows plants to make their own food. The food produced by photosynthesis is the sugar glucose.

In respiration, the glucose is broken down. In this process, energy is released. Plants and animals use this energy to grow, develop, and carry out the activities of living things. *Photosynthesis* is a process that *stores energy*. *Respiration* is a process that *releases energy*.

QUESTIONS	PHOTOSYNTHESIS	RESPIRATION
Where does it happen?	only in cells that contain chlorophyll	in almost all living cells
When does it happen?	only in the presence of sunlight	at all times
What substances are needed?	carbon dioxide and water plus sunlight	glucose and oxygen
What substances are produced?	glucose and oxygen	carbon dioxide and water
What happens to the energy?	stored as glucose	used by cells for various processes

■ *For **Items 1 to 5**, write* T *in the space provided if the statement is true. Write* F *if it is false. If the statement is false, change the underlined words to make the statement true. Write the new word in the space provided.*

_____ 1. Photosynthesis occurs in <u>all cells</u>. ______________________

_____ 2. Respiration occurs in <u>all cells</u>. ______________________

_____ 3. Photosynthesis occurs in the <u>absence</u> of light. ______________________

_____ 4. Respiration requires glucose and <u>carbon dioxide</u>. ______________________

_____ 5. During photosynthesis, energy from the sun is stored in the form of <u>carbon dioxide</u>. ______________________

► *For more information about this topic, see pages 164–165 in Pre-GED Science. Answers begin on page 67.*

Oxygen-Carbon Dioxide Cycle: The Atmosphere Changes

An active reader takes time to preview material before reading it. Quickly skim the passage below. Ask yourself, "What do I already know about this subject?" and "What do I expect to find out?" This will help you stay alert as you read.

Earth's atmosphere was not always as it is today—about 21% oxygen, 78% nitrogen, and 1% other gases. Scientists believe Earth was formed about 4.5 billion years ago. For some time, its atmosphere was mostly made up of two poisonous gases, methane and ammonia. Over many years, the energy of sunlight transformed the gases in the atmosphere. The poisonous gases vanished. The air became rich in nitrogen, hydrogen, carbon dioxide, and water. However, there was still little or no oxygen in the air.

The sun's light split apart some of the water molecules, H_2O, into oxygen and hydrogen. The lightweight hydrogen gases escaped into space. But the oxygen remained behind. Some of the oxygen rose high into the atmosphere and formed a blanket called the *ozone layer*. Today, we know that the ozone layer keeps the sun's harmful ultraviolet energy from reaching Earth.

This ozone layer protected the simple organisms that lived in the oceans. Some of these organisms could perform photosynthesis. As they did so, they poured oxygen into the atmosphere. Later, more oxygen was produced as green plants began to grow on land. By about 600 million years ago, Earth had an atmosphere much like we have today.

■ *Use the information above to answer **Items 1 to 4**.*

1. The early atmosphere of Earth contained what two poisonous gases?

____

2. How did the sun's light help add oxygen to the atmosphere?

____

3. What protects living things from the harmful effects of sunlight?

____

4. What organisms first helped increase the amount of oxygen in Earth's atmosphere?

____

► *For more information about this topic, see page 166 in Pre-GED Science.*
Answers begin on page 67.

Oxygen-Carbon Dioxide Cycle: Unbalancing the Cycle

An active reader takes time to review after reading. Ask yourself, "What did I learn from this reading?" You may find it helpful to take some notes or go back through the reading. Be certain to note any new terms that were not familiar to you. Try this on the passage that follows.

Oxygen and carbon dioxide are gases found in the atmosphere. In nature, the oxygen-carbon dioxide cycle keeps these gases in balance. Huge forests, especially those of the world's tropical countries, take in great amounts of carbon dioxide. They give off great amounts of oxygen. If left alone, these forests would keep these gases in balance.

Unfortunately, the world's forests are being cut down. There are fewer and fewer trees to remove carbon dioxide from the atmosphere. At the same time, artificial sources are putting more and more carbon dioxide into the atmosphere. The primary source of this increase is the burning of fossil fuels. Many scientists fear that this is unbalancing the oxygen-carbon dioxide cycle.

Most of the heat from sunlight escapes back into space. However, carbon dioxide holds some heat near Earth's surface. As the carbon dioxide levels in the atmosphere increase, less heat can escape. Temperatures may rise.

If temperatures rise enough, the polar ice caps will begin to melt. If this occurs, sea levels will rise, flooding coastal lands.

■ *Use the information above to answer **Items 1 to 4**.*

1. As large areas of forests are cut down, what happens to carbon dioxide in the air?

____

2. What other factor is causing an increase in carbon dioxide in the atmosphere?

____

3. What effect does a buildup of carbon dioxide in the air have on Earth's temperature?

____

4. What would happen if the polar ice caps began to melt?

____

► *For more information about this topic, see page 166 in Pre-GED Science.*
Answers begin on page 67.

Habitats: A Planet with Life

Reviewing after reading is especially helpful when there is a lot of new vocabulary. Sometimes you may need to make a vocabulary list or mark new vocabulary as you read.

As far as we know, Earth is the only place where life exists. Life exists in a very thin shell of land, water, and air called the *biosphere*.

The biosphere can be divided into smaller parts called *ecosystems*. An ecosystem can be very large, such as all the oceans of the world. Or it can be small, such as a city garden. Whatever its size, each ecosystem has its own living organisms, nonliving parts, and climate. The nonliving parts include water, soil, rocks, and features such as hills or valleys.

Ecosystems are *self-sustaining*. That is, they contain everything that the organisms in them need to live. For example, in the ocean ecosystem, floating green plants, called *algae*, get energy from sunlight. They use this energy to make their own food. Small shrimp-like animals called *krill* eat the algae. Blue whales, the largest animals to inhabit Earth, eat tons of krill. As long as there is sunlight, and nothing disturbs the balance of life in the ocean, this ecosystem will be self-sustaining.

A *community* consists of all the living things in an ecosystem. Each community is made up of *populations* of different kinds of living things. For example, we can speak of the blue whale population in the ocean. Each population has a *habitat*. A habitat is the place where a particular population lives. For example, the habitat of the ocean algae is the surface of the ocean.

■ *Use information from the passage to answer* ***Items 1 to 3***.

1. What is Earth's biosphere made up of?

____

2. What is the difference between a community and a population?

____

3. What does it mean that an ecosystem is "self-sustaining"?

____

► *For more information about this topic, see pages 168–169 in Pre-GED Science.*
Answers begin on page 67.

LESSON 6 • EXERCISE 2

Food Chains: Reading Diagrams

Some information is easier to understand if it is illustrated rather than written. First scan the whole picture to get an idea of what is shown. Then read any titles or labels to help you understand the illustration.

Food chains show how energy moves from one living organism to others. The beginning of a food chain is usually a green plant. These green plants are called *producers* because they produce their own food energy. The next link in the chain is an animal that eats plants. The final link is usually an animal that eats other animals. Animals are called *consumers*. They must consume, or eat, other organisms for food.

In nature, however, relationships are not that simple. To show the complex relationships among food chains, scientists use food webs like diagram below. The arrows show the direction energy moves through the web. It shows how food chains combine to form webs.

FOOD WEB

■ *Use the information and diagram above to answer* ***Items 1 to 3***.

1. Which organism is a producer?

____

2. Which consumers get energy from eating grasshoppers?

____

3. Which organisms eat both producers and consumers?

____

► *For more information about this topic, see pages 170–171 in Pre-GED Science. Answers begin on page 67.*

Humans and the Environment: Population

Line graphs often are used to show how something changes over time. Read the passage below. Then look at the time scale on the graph. What period of time is covered? Look to see what change is being shown.

Many environmental problems have been caused by people. And the problems have grown worse as Earth's human population has increased. For example, more and more land has been cleared for farms, mines, roads, and cities. This land was once the habitat of countless species of animals and plants. When their habitats were destroyed, some of these animals and plants became extinct. They vanished forever from Earth.

The graph shows how the human population of Earth has changed during the past 500 years. It also shows how the population will change in the future if the present rate of growth continues.

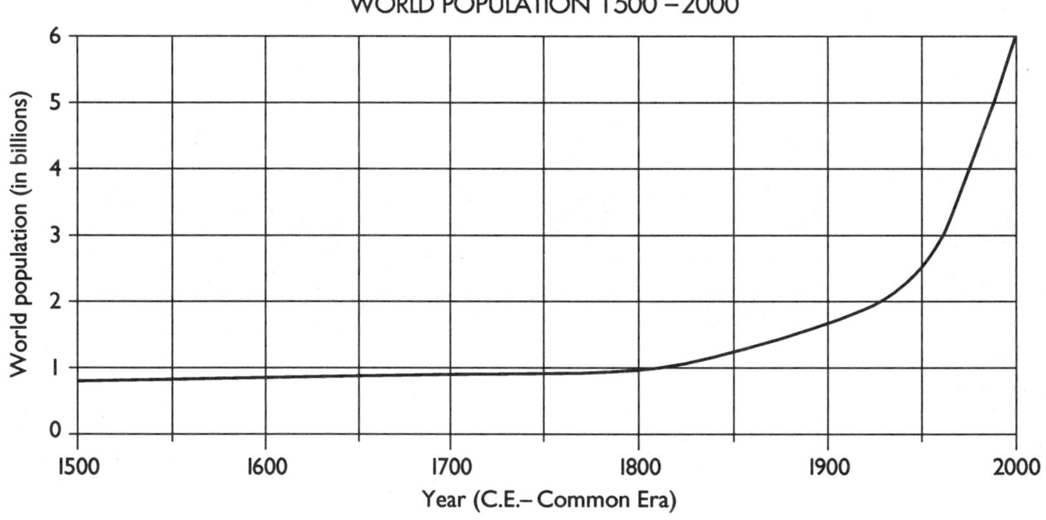

■ *Use the graph and information above to answer* ***Items 1 to 3***.

1. About when did the population reach 1 billion? ______________________

2. About when will the population reach 6 billion? ______________________

3. In which century (100-year period) did the population rise the fastest? ____________

▶ *For more information about this topic, see page 172 in* *Pre-GED Science.*
Answers begin on page 67.

Unit 1 LESSON 6 • EXERCISE 4

Humans and the Environment: Land Use

Bar graphs are used to compare amounts, not to show a continuous change. Read the passage below. Then look at the *x-axis* and the *y-axis* to see what is being shown on the graph. One axis will show what is being compared. The other will tell how it is being compared.

Human activities have changed the way people use land. The country of Zimbabwe in Africa is a good example of how the use of land changes in a developing country. In the 1990s, 3,500 acres of land were cleared every day to build factories, buildings, and roads. Little by little, the rural, or country, land was changed into urban, or city, land.

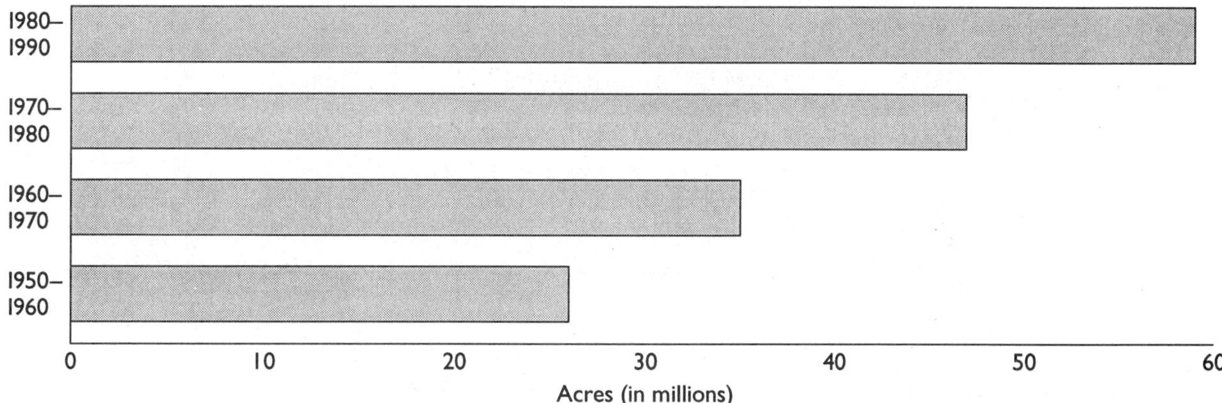

■ *Use the graph and the information above to answer* ***Items 1 to 4***.

1. Between what years were 35 million acres of rural land changed into urban land?

2. In what ten-year period was the least amount turned into urban land?

3. In what ten-year period was the most amount turned into urban land?

4. What do you think the bar for 1990–2000 will look like? Why?

► *For more information about this topic, see page 172 in* Pre-GED Science.
Answers begin on page 67.

Unit 1: Optional Writing Activity

■ *Choose one of the writing suggestions below. Write a paragraph on that topic.*

1. Review Lesson 1, Exercise 4, *Specialized Cells in Humans* on page 4. Summarize how HIV causes AIDS and why AIDS can be fatal.

2. Review Lesson 3, Exercise 2, *Fetal Development* on page 11. Describe how the placenta helps support the life of a fetus.

3. Review Lesson 6, Exercise 3, *Humans and the Environment* on page 26. Write an opinion about whether or not you think an increase in the world's human population is a serious concern.

► *Answers begin on page 68.*

Unit 2 LESSON 7 • EXERCISE 1

The Structure of Earth: Earthquakes

Finding cause-and-effect relationships is an important skill. Remember that causes are reasons. Effects are results. As you read, try to identify relationships that are cause-and-effect.

Earth is made up of three layers: the crust, the mantle, and the core. You stand on the top layer of Earth, its crust. Every now and then, parts of the crust suddenly move. This movement is called an *earthquake*. It happens somewhere along a crack in the crust called a *fault*.

The best known fault in the United States is in California. The San Andreas fault is 960 kilometers long (almost 600 miles) and 32 kilometers deep. This fault begins in the Pacific Ocean north of San Francisco and runs down through the state to the Gulf of California. Movement along this fault caused a major earthquake in San Francisco in 1989.

The land on the west side of the San Andreas fault is creeping north very slowly. The land on the east side is moving equally slowly to the south. But the movement isn't steady. It comes in small skips. When one of these skips happens, ripples of energy spread out across the land. These ripples of energy are called *earthquake waves*. Like water waves, earthquake waves make things move. If the movement is great enough, houses tumble down, roads crack, and bridges collapse.

■ *Use the information in the passage above to answer* ***Items 1 to 4***.

1. In which of Earth's layers do earthquakes occur?

____

2. What is a fault?

____

3. What is the cause of an earthquake?

____

4. What are two possible effects of earthquakes?

____

____

► *For more information about this topic, see pages 176–177 in Pre-GED Science. Answers begin on page 68.*

The Structure of Earth: The Ocean Floor

Applying written information from a passage to a diagram is a skill often used in science. Read the passage carefully, noting any *italicized* words. Read any titles or labels on the diagram. Then see how the two are related.

The land beneath the oceans has deeper canyons, larger plains, and taller mountains than the land on the continents. For example, the island of Hawaii is at the top of an underwater mountain. This mountain rises more than 9,600 meters above the ocean floor. By contrast, Mt. Everest is about 9,000 meters high. In science, the term *mountain* usually refers to land formations found above sea level. Otherwise, Hawaii would be the tallest mountain on Earth!

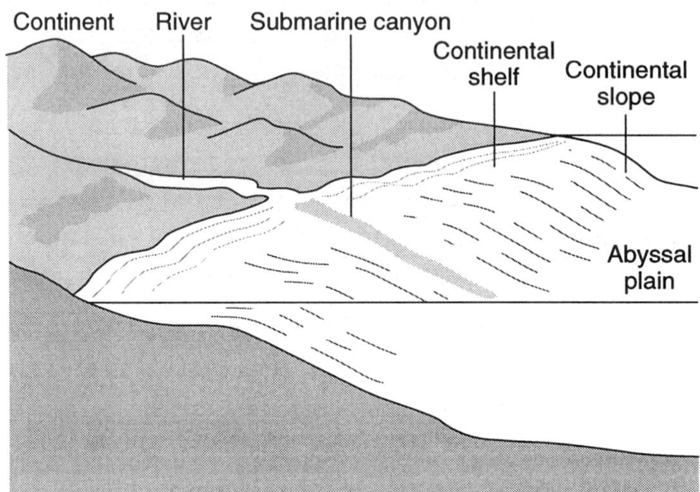

The land on the continents extends out from the shore line. This land, called the *continental shelf*, may be as little as 10 kilometers or as much as 1,200 kilometers wide. Where it ends, the land drops steeply to a depth of 4 to 5 kilometers. This area of rapid change is called the *continental slope*. The flat area at the bottom of the continental slope is referred to as the *abyssal plain*.

■ *For **Items 1 to 3**, use one of the italicized words in the passage to name the steps in the drawing above.*

1. ______________________ 3. ______________________

2. ______________________

■ *Use the information in the passage above to answer **Item 4**.*

4. Why might it be true to say that the people on the island of Hawaii live in the mountains?

____

► *For more information about this topic, see pages 176–177 in Pre-GED Science. Answers begin on page 68.*

LESSON 7 • EXERCISE 3

Earth's Resources: Using Fresh Water

Sometimes by studying a graph, we are able to make inferences. As with all graphs that have an x-axis and a y-axis, be certain you know what information is shown on each axis before you try to make any inferences.

Fresh water is one of our most precious natural resources. We need it for drinking, washing, and cooking. Although water is a renewable resource, periods of drought have often reduced its supply. In addition, many parts of the world do not get enough rainfall to supply human needs.

It is very important to conserve fresh water. But to do this, people need to know how much they use and how much they waste. The graph shows how the average American family of four uses fresh water.

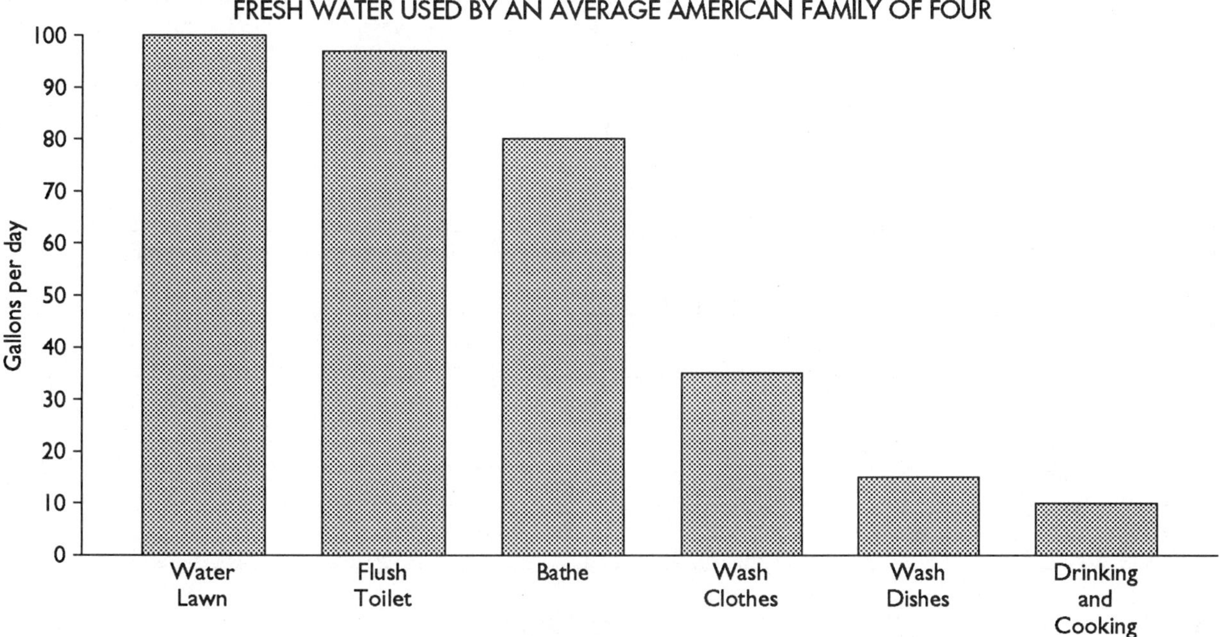

■ *Use information in the graph to answer **Items 1 to 5**.*

1. How many gallons of water a day are used for bathing? ______________________________

2. Which activity uses up the least amount of water per day? ______________________________

3. Which activity uses up about 100 gallons of water each day? ______________________________

4. If you lived in a city apartment, what would be your greatest use of water? ______________

5. Can you infer one way to conserve fresh water? ______________________________

► *For more information about this topic, see pages 178–179 in* <u>*Pre-GED Science*</u>*.*
Answers begin on page 68.

Earth's Resources: Minerals

Recall the questions that are useful for finding details—*who? what? where? when? why?* and *how?* As you read, ask yourself these questions. Picture the answers in your mind.

Minerals are nonrenewable resources. There are only limited amounts of minerals in the ground. Nevertheless, very large amounts have been removed from the ground to make all sorts of products. The table below shows some mineral resources and their uses.

MINERAL RESOURCES	
Mineral	**Some Uses**
aluminum	to make airplanes, cars, containers for beverages; in toothpaste, aluminum wrap
copper	in electrical wiring, plumbing (pipes), pots and pans; to make jewelry, brass, and bronze
gold	to make jewelry; in dental work, coins
mercury	in thermometers and barometers, street lights, thermostats
nickel	to make steel, coins; in metal plating
silicon	electronic devices; solar batteries, silicones, ceramics
silver	to make jewelry, tableware; in photographic film, coins
tungsten	in light bulbs, dental work

■ *Use the passage and chart to answer* ***Items 1 to 3.***

1. Who uses minerals?

____

2. What two minerals are involved when you switch on a light?

____

3. Where are copper, silver, and gold all used?

____

► *For more information about this topic, see pages 178–179 in* *Pre-GED Science.*
Answers begin on page 68.

LESSON 7 • EXERCISE 5

The Greenhouse Effect: Carbon Dioxide Blanket

Identifying implications, or reading between the lines, is an important skill in science. With this skill, you can use information given to understand something that is not actually stated.

As carbon dioxide increases, the average temperature of Earth goes up. This is caused by the greenhouse effect, or the trapping of the sun's heat near Earth's surface. Carbon dioxide acts as a blanket that holds this heat in. If there is more carbon dioxide, more heat is trapped, and the Earth gets warmer.

Scientists tell us that even a small increase in Earth's average temperature can have serious effects. Increased temperature may turn farmland into desert. Polar ice caps may begin to melt, causing other farmland to be flooded. The graph below shows the carbon dioxide concentration in the air from 1956 to 1990.

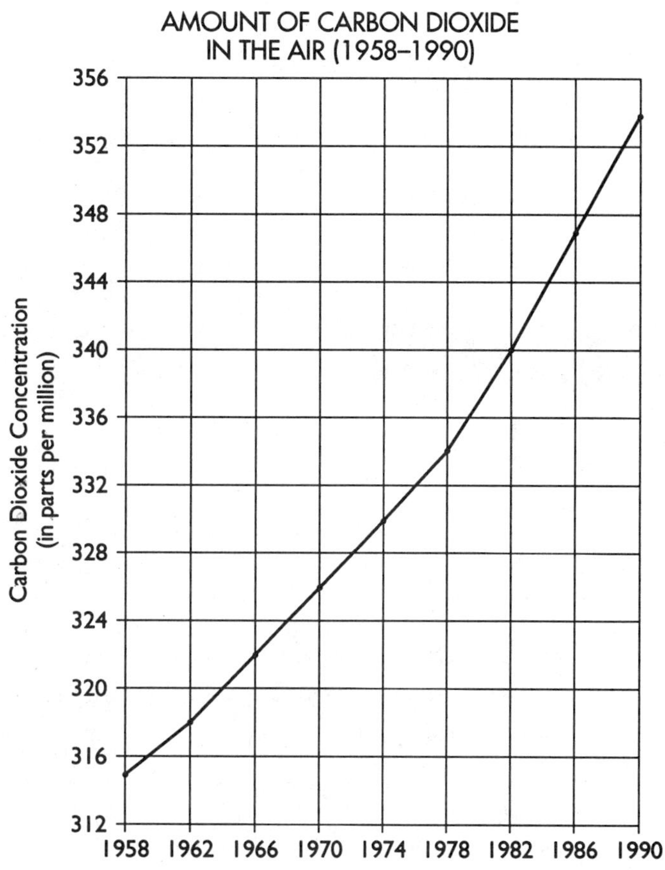

■ *Use the graph and information above to answer* ***Items 1 to 4.***

1. What was the concentration of carbon dioxide in the air in 1958? ________________

2. About when did the carbon dioxide concentration reach 335 parts per million? _______

3. What was the concentration of carbon dioxide in the air in 1990? ________________

4. What does the graph imply about Earth's average temperature between 1958 and 1990? (Hint: How does the carbon dioxide concentration change during this period? What would result from this change?)

____

► *For more information about this topic, see page180 in* *Pre-GED Science.*
Answers begin on page 68.

Unit 2 LESSON 8 • EXERCISE 1

The Earth-Sun Relationship: Changing Daylight

Reading a table is an important skill in everyday life as well as in science. Remember that the columns, running up and down, show what is being compared. The rows, running across, list specific data.

When the northern hemisphere is tilted toward the sun, it is warmer in the northern hemisphere. This change in position causes the length of daylight and the season to change. The table below shows the change in the number of daylight hours at Seward, Alaska during the year.

Earth is *not* tilted away from or toward the sun on the first day of spring or autumn. In the northern hemisphere, the first day of spring is near March 21. The first day of autumn occurs near September 22.

LENGTH OF DAYLIGHT IN SEWARD, ALASKA

Date	Daylight (hours:minutes)	Date	Daylight (hours:minutes)
March 20	12:12	September 20	12:37
June 20	18:52	December 20	5:53

■ *For **Items 1 to 4**, choose the one word or phrase from the list below that correctly completes each of the following statements. Note that there is one extra answer choice.*

winter summer fall and spring increases decreases

1. Daylight hours are longest during ____________________.
2. Daylight hours are shortest during ____________________.
3. Days and nights are about equal on the first day of ____________________.
4. From spring to summer, daylight ____________________.

► *For more information about this topic, see pages 182–183 in* Pre-GED Science. *Answers begin on page 68.*

The Earth-Sun Relationship: Phases of the Moon

Many diagrams in science show a sequence of events. As with any diagram, first read the title and labels. Look closely to see if there is a numbered or lettered sequence or arrows that show what change is taking place.

Our moon revolves around Earth. We are able to see the moon because sunlight is reflected off it. What we call moonlight is actually reflected sunlight. Sunlight illuminates, or lights, only half of the moon. Because of the movement of Earth and the moon, we do not always see the entire illuminated half of the moon. As a result, the shape of the moon appears to change. The different shapes that we see are called the *phases* of the moon. The complete cycle takes about 29.5 days. The diagram shows what these phases are called.

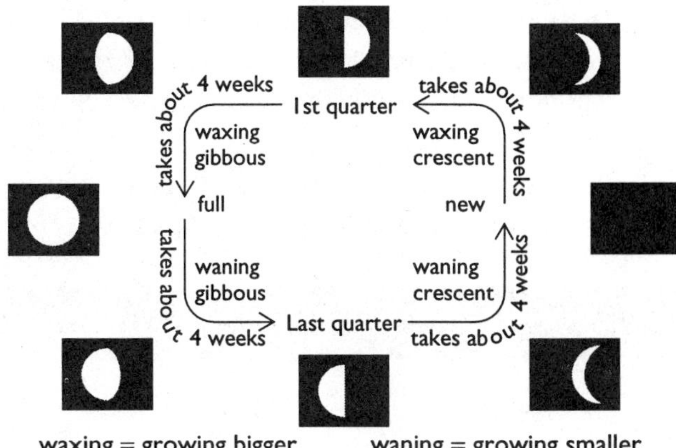

■ *Use the diagram and the passage above to answer* ***Items 1 to 4****.*

1. What causes the stages that we call the phases of the moon?

____

2. When we see a "full" moon, how much of its surface is actually lighted?

____

3. Which phase follows the last quarter stage?

____

4. What stage do you think occurs when the moon is between Earth and the sun? Explain your choice.

____

► *For more information about this topic, see pages 182–183 in* Pre-GED Science. *Answers begin on page 69.*

 LESSON 8 • EXERCISE 3

The Solar System: Planets

Graphics, including tables, can help you compare and contrast items. Comparisons tell how things are alike. Contrasts tell how things are different.

The solar system is made up of nine planets. A planet's year is the time it takes for a planet to orbit the sun. The length of each planet's year depends on the planet's distance from the sun. The closer a planet is to the sun, the shorter its year. Some are tens of millions of miles from the sun, while others are billions of miles from the sun. Some have long days, while others have short days. Planets differ in other ways. Their diameters, or distances across, vary.

THE PLANETS

Planet	Diameter (km)	Average Distance from Sun (km)	Length of Day (Earth Time)
Mercury	4,880	58,000,000	58.7 days
Venus	12,104	108,000,000	243 days
Earth	12,756	150,000,000	24 hrs
Mars	6,794	228,000,000	24.5 hrs
Jupiter	142,700	778,000,000	9.9 hrs
Saturn	120,000	1,427,000,000	10.5 hrs
Uranus	50,800	2,869,000,000	17 hrs
Neptune	48,600	4,486,000,000	18.5 hrs
Pluto	4,000	5,890,000,000	9.5 hrs

■ *Use the table above to answer **Items 1 to 4**.*

1. Which is the second largest planet? ____________________

2. Which is farther from the sun, Earth or Mercury? ____________________

3. Which planet has the longest day? ____________________

4. Which planet's day is closest in length to Earth's day? ____________________

► *For more information about this topic, see pages 184–185 in Pre-GED Science. Answers begin on page 69.*

 LESSON 8 • EXERCISE 4

The Universe: The Milky Way

Often, it is necessary to classify information, or put it into groups. First note similarities or differences among items. Then you can make a decision about how to classify items. Often size is used to classify items.

When you look up into the night sky, you can see many stars. A family of stars is called a *galaxy*. Our galaxy is the Milky Way galaxy. Our sun, which is a *star*, belongs to this galaxy. So does Earth and all the other planets and objects of our solar system. Across the middle of the sky you will see a concentration of stars. This is the thickest part of the Milky Way. This star-studded belt is what gives our galaxy its name. It looks as if someone had poured a streak of milk across the sky.

There are between 100 and 200 billion stars in the Milky Way. This galaxy is so huge that even if you could travel 40,000 kilometers an hour, it still would take you more than 20 trillion years to cross the galaxy! Although the Milky Way galaxy is one of the largest galaxies in space, it is tiny compared to the size of the *universe* itself. Scientists estimate that the universe is at least 150,000 times larger than the Milky Way. So, if the Milky Way were the size of a nickel, the universe would be the size of a circle almost two miles wide.

■ *For **Items 1 to 7**, use one of the following terms to complete each sentence. Terms may be used more than once.*

star galaxy universe

1. The Milky Way is a ________________.
2. Our sun is a ________________.
3. Our solar system is part of a ________________ and a ________________.
4. A galaxy is larger than a ________________.
5. The Milky Way is smaller than the ________________.
6. The ________________ contains more stars than a ________________ does.
7. It would take longer to travel across a ________________ than a ________________.

► *For more information about this topic, see pages 186 in Pre-GED Science. Answers begin on page 69.*

The Universe: Variations Among Stars

Tables can show relationships that may not be seen as easily in written text. From studying these relationships, it may be possible to make predictions.

Although the stars in the sky may look just about the same, they can be very different from one another. Some are very young, while some are middle aged. Still others are rather old. Stars also vary in temperature and color. Temperatures may range from 35,000°C to 3,000°C, which is cool for a star. Their colors may range from blue or blue-white to red.

Stars also vary in size. Some are much larger than our star, the sun. One such star is the red super giant Antares. Its diameter is about 400,000,000 kilometers. If the sun were that size, it would swallow up Mercury, Venus, and Earth.

CHARACTERISTICS OF STARS

Star	Color	Surface Temperature (C°)
Algol	blue-white	35,000
Sirius	white	10,000
Ploaris	yellowish	7,500
Sun	yellow	6,000
Aldebaran	red-orange	4,000
Antares	red	3,000

■ *Complete **Items 1 to 5** by filling in the blanks.*

1. The colors of the hottest stars are ________________.
2. Antares, the coolest star on the chart, is ________________ in color.
3. Although not shown in the table, the star Vega has a temperature of about 10,000°C. So, Vega should resemble the star ________________.
4. If a star had a temperature of 6800°C, you could expect its color to be ________________.
5. You can predict the temperature of a star from its ________________.

▶ *For more information about this topic, see page 186 in Pre-GED Science. Answers begin on page 69.*

Unit 2: Optional Writing Activity

■ *Choose one of the writing suggestions below. Write a paragraph on that topic.*

1. Review Lesson 7, Exercise 3, *Earth's Resources* on page 31. Suggest two more ways everyone could conserve fresh water. Could you, personally, use these three methods over a long period of time?

2. Review Lesson 7, Exercise 4, *Earth's Resources* on page 32. Think of three items on the mineral chart that you use every day. What mineral resources were needed to make these items? How would your life be different if you didn't have these items?

3. Review Lesson 8, Exercise 1, *The Earth Sun Relationship* on page 34. During warmer weather, most states go on daylight savings time. Clocks are set so that there are more hours of sunlight in the evening. Why do you think people want those daylight hours in the evening, rather than in the morning?

► *Answers begin on page 69.*

Unit 3 LESSON 9 • EXERCISE 1

Atoms and Molecules: Bits and Pieces

One way to *classify* is to group things together according to something the items have in common. Each group is given a name to identify it.

Everything you see, from a speck of dust to the moon, is made of matter. You could say that matter is "stuff." The stuff, or substance, may be visible, like clouds, a grain of sand, a chair, or a sweater. Or it may be invisible, like air. All matter is made up of smaller particles called *atoms*. Atoms are so small that you cannot see a single atom. If you look at a gold ring, you are looking at billions of atoms of gold crowded together.

Atoms are themselves made up of smaller pieces. These pieces are so small that they are invisible—even through the most powerful microscope! These pieces of matter are called *subatomic particles*. Subatomic means smaller than an atom. Subatomic particles include *protons*, *neutrons*, and *electrons*.

When atoms combine, they make larger pieces of matter called *molecules*. Water is made up of hydrogen and oxygen. Atoms of hydrogen and oxygen are joined together to form a molecule of water. When billions of water molecules come together, you see a drop of rain!

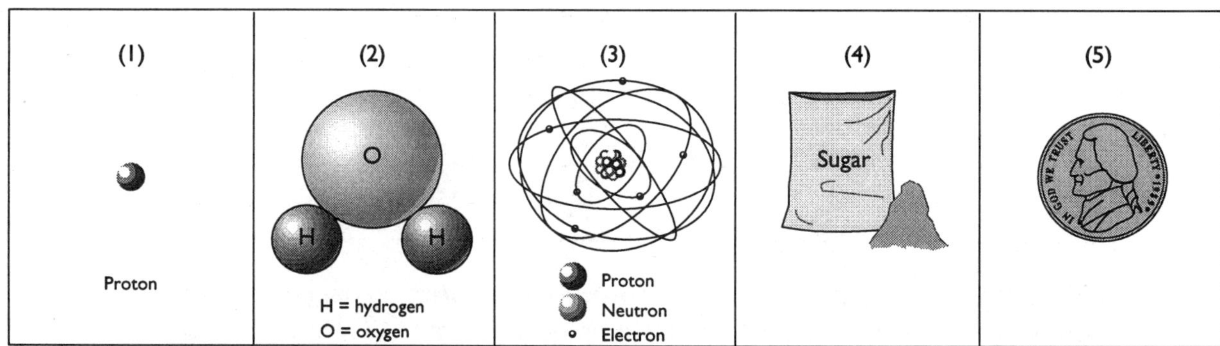

■ *For* ***Items 1 to 5,*** *choose one of the following terms to classify the matter in each of the drawings above. Answers may be used more than once.*

subatomic particle   atom   large group of atoms or molecules   molecule

1. ______________________
2. ______________________
3. ______________________
4. ______________________
5. ______________________

► *For more information about this topic, see pages 190–191 in Pre-GED Science. Answers begin on page 69.*

Elements: What Makes Elements Different?

A topic sentence tells the main idea of a paragraph. Locating the topic sentence will help you understand what you read. You then can express the main idea of a paragraph or passage in your own words.

Gold, iodine, oxygen, and carbon are very different elements. They are only four of the 109 elements known to scientists. What makes one element different from another? The answer is found in its atoms. An atom of one element is unlike an atom of any other element.

What exactly makes the atoms of the elements different? The atoms of each different element contain a different number of protons. For example, gold atoms have 79 protons; iodine has 53; oxygen has 8; and carbon has 6.

This difference gives each element its individual characteristics. Gold is a shiny metal. Iodine is a dark, poisonous solid. Oxygen is a tasteless, colorless gas. The element carbon usually takes the form of a sooty, black material. Other elements have different characteristics also. Mercury is a thick, silvery liquid. Barium is a soft, silvery-white solid.

■ *Use the information above to answer* ***Items 1 to 4***.

1. What is the topic sentence in the first paragraph?

____

____

2. What is the topic sentence of the second paragraph?

____

____

3. What is the main idea found in the last paragraph?

____

____

4. In your own words, what is the main idea of the passage?

____

► *For more information about this topic, see pages 192–193 in* Pre-GED Science.
Answers begin on page 69.

Elements: Telling Elements Apart

Sometimes information can be taken directly from a table. At other times you need to use the information to infer, or figure out, something else.

Every element has a set of physical properties. Examples of physical properties include color, density, boiling point, and melting point. The *boiling point* is the temperature at which a substance changes from a liquid to a gas. The *melting point* is the temperature at which a solid turns to a liquid. *Density* tells how much mass is contained in a substance. For example, suppose you had cube-shaped samples of various elements. Each cube is exactly the same size. One sample has a mass of 1 gram. Another sample has a mass of 1.8 grams. Thus, the second sample is more dense. The higher the number, the more dense the matter.

PROPERTIES OF SOME ELEMENTS

Element	Color	Density (g/cm^3)	Melting Point (°C)	Boiling Point (°C)
sodium	silvery	0.97	97.5	880
iron	silver-gray	7.86	1535	3000
zinc	bluish-white	7.14	419.5	907
silver	silver	10.5	960.8	1950
gold	yellow	19.3	1063.0	2600
platinum	silver	21.45	1773.5	4300
mercury	silver	13.55	-38.9	357

■ *Use the table and information above to answer* ***Items 1 to 4***.

1. Of all the silver-colored metals, which is the most dense? ________________

2. What is the boiling point of zinc? ________________

3. Room temperature is about 21°C. Which of the metals is a liquid at room temperature?

   ____

4. If you had samples of all these elements on a table, which could you identify most easily? How?

   ____

▶ *For more information about this topic, see pages 192–193 in Pre-GED Science. Answers begin on page 70.*

The Periodic Table: Groups of Elements

Finding details in a passage or chart is a very important skill. A *detail* is a bit of information that supports a main idea.

The elements in the periodic table are placed in columns running up and down. Elements in the same column make up a *group*. Groups of elements are similar. They combine with other substances in similar ways. For example, Group 18 is on the far right side of the periodic table.All the elements in this group are gases. These elements do not combine easily with other elements. We say that they are *chemically stable*. You are probably familiar with the Group 18 element neon. It is used in making colorful lights for signs.

The elements in the column just to the left are much different. In Group 17, there are elements that are gases and solids. One element is a liquid. In Group 17, the element at the top combines most easily with other atoms. The element at the bottom combines least easily. As you move down both columns, each element has larger atoms than the element just above it.

GROUPS 17 AND 18 OF THE PERIODIC TABLE

17	18
	2 4.0 **He** Helium
9 18.9 **F** Fluorine	10 20.1 **Ne** Neon
17 35.4 **Cl** Chlorine	18 39.9 **Ar** Argon
35 79.9 **Br** Bromine	36 83.8 **Kr** Krypton
53 126.9 **I** Iodine	54 131.3 **Xe** Xenon
85 (210)* **At** Astatine	86 (222)* **Rn** Radon

□ Liquid at room temperature
□ Gas at room temperature
□ Solid at room temperature

■ *Use the information in the passage and from the periodic table to answer **Items 1 to 5**.*

1. In Group 18, which element has the largest atoms?

____

2. In Group 17, which element combines more easily with other elements, chlorine or iodine? ______________________

3. Which elements in Group 17 are solids?

____

4. In what way are groups of elements similar?

____

____

5. What does the term "chemically stable" mean?

____

▶ *For more information about this topic, see page 194 in Pre-GED Science. Answers begin on page 70.*

Unit 3 LESSON 10 • EXERCISE 1

Three States of Matter: Gas Laws

Sometimes you have to apply information to a new situation. Scan the information quickly. Think about what you already know about the topic. Then read the passage carefully.

A gas takes the shape of whatever container it is in. It fills the entire container. We call this its *volume*. Volume is the amount of space something takes up.

The volume of a gas is affected by *temperature*. Suppose you blew up a balloon in a cool room. Then you went outside where it was quite warm. The gas inside the balloon would expand, or push out more against the balloon's sides. If you were to place the same balloon in a freezer, the gas inside would contract, or take up less space. As temperature increases, the volume of a gas increases. As temperature decreases, the volume decreases.

The volume of a gas also is affected by *pressure*. If you press down hard on a gas, its volume becomes smaller. With gases, if pressure is increased, volume is decreased. If pressure is decreased, volume is increased. You may have heard of some machines, such as drills, which use compressed air. The air has been compressed, or squeezed, into a small space. When this air is released, it comes out with great force.

■ *Use the information above to fill in the blanks in **Items 1 to 4**.*

1. The volume of a gas is affected by both ________________ and ________________.
2. Volume is the amount of ________________ a substance takes up.
3. If the pressure on a container of gas is decreased, its volume will ________________.
4. If the temperature of a gas is decreased, the volume will ________________.

■ *Apply what you have learned to answer **Items 5 and 6**.*

5. If you continue to blow up a balloon, at some point it will burst. Which is involved, temperature or pressure? Explain.

____

6. Think about what happens to air in cold weather. Then explain why tires need more air in them in winter than in summer.

____

► *For more information about this topic, see pages 196–197 in Pre-GED Science. Answers begin on page 70.*

Unit 3 LESSON 10 • EXERCISE 2

Three States of Matter: Boiling Point and Air Pressure

When a passage includes a graph, be certain you understand what the graph represents. Read its title. Look to see what the *x*- and the *y-axis* represent. As you read, see how the graph is related to the text.

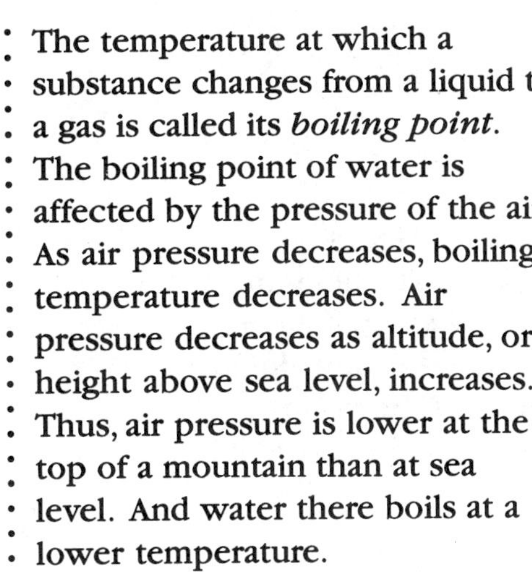

The temperature at which a substance changes from a liquid to a gas is called its *boiling point.* The boiling point of water is affected by the pressure of the air. As air pressure decreases, boiling temperature decreases. Air pressure decreases as altitude, or height above sea level, increases. Thus, air pressure is lower at the top of a mountain than at sea level. And water there boils at a lower temperature.

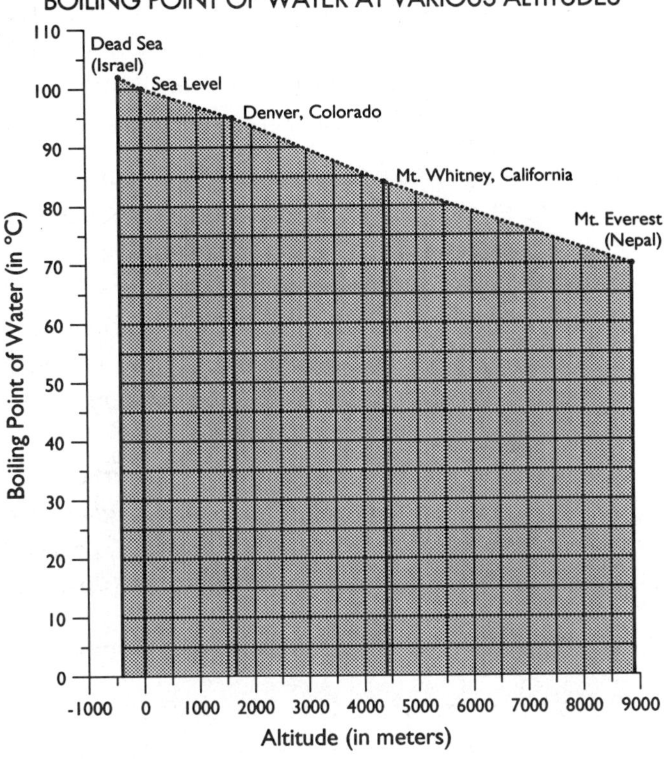

■ *Use the passage and the graph to answer **Items 1 to 3**.*

1. Where does water boil at 95°C? ____________________

2. At about what temperature does water boil at the Dead Sea? ____________________

3. At about what altitude does water boil at 70°C? ____________________

■ *For **Items 4 to 7**,* circle the correct term in parentheses.

4. Look at the dotted line on the graph. It shows an (increase/decrease) in boiling point as altitude increases.

5. As altitude increases, air pressure (increases/decreases).

6. An increase in air pressure causes an (increase/decrease) in boiling point.

7. Water boils at a (higher/lower) temperature in Denver than on Mt. Everest.

▶ *For more information about this topic, see pages 196–197 in Pre-GED Science. Answers begin on page 70.*

Physical and Chemical Changes: Changes All Around You

Classification is the sorting of things with common qualities into groups. It is an important skill in science. Remember that items, ideas, and examples often can be sorted in many different ways.

Every day you see both physical and chemical changes. Remember that in a physical change, no new substance is formed. The physical appearance of the substance simply changes. In a chemical change, however, a completely new substance is formed.

If sugar crystals are ground up, powdered sugar results. This is a physical change. Powdered sugar is still sugar. It just looks different from sugar crystals. However, if the sugar were heated in a pan on a burner, it would begin to change. First the sugar would become liquid. Then it would bubble and turn pale brown. As the sugar heats up, it would become darker and darker. If left on the burner, the sugar would begin to burn. A chemical change would occur.

■ *Classify the following as either physical or chemical changes. Ask yourself if a new substance has formed or if the substance has only changed in form or physical appearance. For* ***Items 1 to 10****, place a* C *for chemical change, or a* P *for physical change in the blank before each item.*

_____ **1.** A piece of meat is ground in a meat grinder.

_____ **2.** A piece of ice is chipped off a larger block of ice.

_____ **3.** A marshmallow is toasted over an open flame.

_____ **4.** An icicle melts.

_____ **5.** A cup is dropped and broken.

_____ **6.** Milk left out of the refrigerator turns sour.

_____ **7.** Grape juice changes into wine.

_____ **8.** Yarn is woven into cloth.

_____ **9.** Water in a pipe freezes and the pipe breaks.

_____ **10.** A piece of fruit begins to rot.

▶ *For more information about this topic, see pages 198–199 in* Pre-GED Science. *Answers begin on page 70.*

Unit 3 LESSON 10 • EXERCISE 4

Chemical Reactions: Speeding Up and Slowing Down Reactions

Frequently science passages contain words that you don't know. You have to figure out what the words mean from how they are used. This is called *using context clues*.

Not all chemical reactions take place at the same rate. Some chemical reactions take place very quickly. For example, a match bursts into flames. Other reactions are slow like a car becoming rusty.

Catalysts are substances that speed up the rate of a chemical reaction. However, the catalyst itself is not changed in the reaction. In the human body, many enzymes act as catalysts. They allow important chemical reactions to take place in the body's cells. Without these enzymes, the necessary reactions would take too long. Catalysts also are found in other, quite different, reactions. A catalyst is used to break down toxic exhaust fumes from cars. The *catalytic converter* in a car changes deadly carbon monoxide into carbon dioxide.

Inhibitors slow down chemical reactions. The food industry uses inhibitors to keep food fresh. BHT and BHA are two common food preservatives.

■ *For **Items 1 to 6**, match the underlined word from the passage with the correct definition.*

______	**1.** rate	**a.**	chemical substances found in the body
______	**2.** burst	**b.**	substances that prevent or slow down decay
______	**3.** enzymes	**c.**	a sudden, strong action
______	**4.** toxic	**d.**	poisonous
______	**5.** exhaust fumes	**e.**	gases from an exhaust pipe
______	**6.** preservatives	**f.**	speed

■ *Use the passage above to answer **Items 7 and 8**.*

7. What is the difference between a catalyst and an inhibitor?

____

8. What do catalysts do in your body? Why are they so important?

____

▶ *For more information about this topic, see page 200 in Pre-GED Science. Answers begin on page 70.*

Unit 3 LESSON 11 • EXERCISE 1

Compounds: Amazing Changes

Finding details involves asking the right questions. Recall the six W and H questions: *who? what? where? when? why?* and *how?* Asking these questions will help you find important details in a passage.

Chemical compounds are made up of the atoms of two or more different elements. Individual atoms are linked together to make the compound. The compound, however, is different from its individual parts.

Sodium is a soft, bright, silver-colored metal. Its symbol is *Na*. It combines violently with water. Sodium is so reactive that it is never found alone in nature. It is always combined with other substances. Chlorine is a greenish-yellow poisonous gas. Its symbol is *Cl*. If you were to breathe this gas, you would damage your lungs seriously. Chlorine also is very reactive. It, too, is not found free in nature.

If you combine sodium with chlorine, you get a compound. The formula of this compound is *NaCl*. The compound is called sodium chloride. Sodium chloride is a white solid. If you look at it under a microscope, you can see that it is made of crystals. These crystals are not poisonous. You know this compound as table salt.

Table sugar is made of three elements: carbon, oxygen, and hydrogen. TNT, an explosive, is made of four elements: carbon, oxygen, hydrogen, and nitrogen. Carbon usually exists as a black solid. Oxygen, hydrogen, and nitrogen are colorless gases. However, table sugar is a sweet-tasting, white solid. And TNT is a colorless solid. Clearly, compounds are not much like the elements that form them.

■ *For* ***Items 1 to 4****, fill in the blank with the correct answer.*

1. A compound is made up of ____________________ elements.
2. Sodium is a ____________________ metal.
3. Chlorine is a ____________________ gas.
4. Sodium chloride is a ____________________.

■ *For* ***Items 5 to 8****, circle the correct term in parentheses.*

5. Sodium is a(n) (element/compound).
6. Table sugar is a(n) (element/compound).
7. A compound is (similar to/unlike) the elements of which it is made.
8. Elements in a compound are (linked together/not linked).

► *For more information about this topic, see pages 202–203 in* *Pre-GED Science.*
Answers begin on page 70.

Compounds: Hydrocarbons

Some information is much easier to understand if it is illustrated rather than written. When you look at illustrated, or visual, information, first scan the overall picture. Then read the title or any labels on the illustration. If more than one item is shown, look to see how the items are alike or different.

A compound made only of the elements carbon and hydrogen is called a *hydrocarbon*. Hydrocarbons are an extremely important group of compounds. More than 90% of all energy sources used are hydrocarbons. Some common hydrocarbons are: gasoline, kerosene, and methane.

Crude oil (out of the ground) is a mixture of many different hydrocarbons. The oil is taken to a refinery where the different hydrocarbons are separated. Each hydrocarbon has a different boiling point. The refinery uses the different boiling points as a way to separate the hydrocarbons.

The illustration below shows three hydrocarbons: methane, ethane, and propane.

Methane B.P. = -161.5°C   Ethane B.P. = -88.3°C

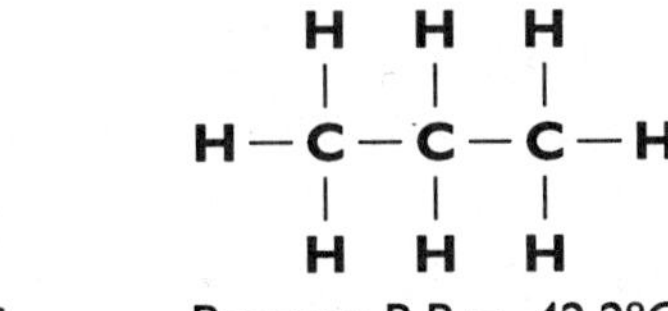

Propane B.P. = -42.2°C

Key: H = hydrogen C = carbon B.P. = boiling point — = link between atoms

■ *Use the information and illustration to answer* ***Items 1 to 5.***

1. What two elements are found in hydrocarbons? ____________________

2. Look at the illustration of methane. How many carbon atoms are in methane? How many hydrogen atoms?

____

3. What is the boiling point of ethane? ____________________

4. How many carbon atoms are in propane? ____________________

5. How does the boiling point change as the number of carbon atoms in these three compounds increase?

____

▶ *For more information about this topic, see pages 202–203 in* *Pre-GED Science.* *Answers begin on page 70.*

Mixtures: Telling Compounds and Mixtures Apart

When you classify, you group things. Usually things are grouped according to what they have in common. Before you classify an item, be certain you know the characteristics of the groups.

A compound is a combination of the atoms of two or more elements. Only a chemical reaction will separate a compound into its individual elements. Scientists call this process *separating materials by chemical means*. A mixture, on the other hand, is made up of different compounds or other materials. You do not have to create a chemical reaction to separate a mixture into its parts.

Suppose you are given a glass of what looks like brown water. Are you holding a mixture or a compound? How do you tell? If you can separate the substances in the glass without using a chemical reaction, you have a mixture. Scientists call this process *separating materials by physical means*.

Filtration is one way to separate a liquid mixture by physical means. Suppose you poured the brown liquid through filter paper. The little pieces of matter mixed in the liquid would stay on the filter paper. The liquid that passed through the filter paper would appear clear or almost clear. Since you have separated the solid from the liquid by physical means, you know you had a mixture. Coffee filters will keep larger coffee grounds from passing through. However, they are not fine enough to keep small particles out. Some filters used in industry will filter out extremely fine particles.

■ *Choose one of the following terms to complete* ***Items 1 to 6****. Terms can be used more than once.*

Compounds   Mixtures   Compounds and mixtures

1. ______________________ can be separated into parts only by chemical means.
2. ______________________ can be separated into parts by filtration.
3. ______________________ are made up of atoms.
4. ______________________ have atoms that are held together tightly.
5. ______________________ can be separated by physical means.
6. ______________________ may be liquids containing particles that can be filtered out.

► *For more information about this topic, see pages 204–205 in Pre-GED Science. Answers begin on page 71.*

Mixtures: Checking Compounds and Mixtures

Illustrations as well as words can be compared and contrasted. When two or more similar illustrations are shown, read any text or labels. Ask yourself how the illustrations are alike. Then look for differences.

Some compounds and mixtures can look exactly alike. What's more, if you pass them through a sheet of filter paper, they still look alike. For example, suppose you have two glasses of clear liquid. One is a glass of pure water. The other is a glass of salty water. They look exactly alike. The illustration below shows a physical process used to separate liquids. In Step B, the liquids are heated to allow the water to evaporate more quickly. The results in Step C tell you if the liquid was a mixture or not.

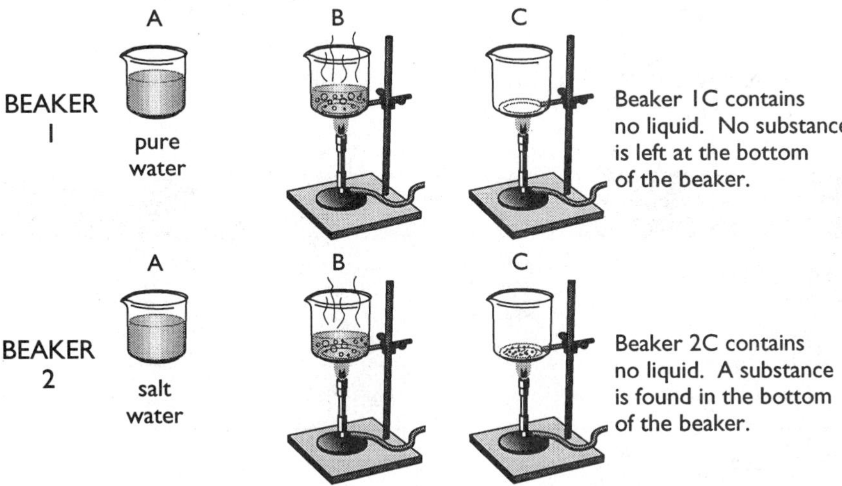

■ *Use the illustration and passage to answer **Items 1 to 3**.*

1. What is similar about the liquids in Step A? What is different?

____

2. What is the difference between the beakers in Step C?

____

3. Which beaker clearly contained a mixture? Explain your answer.

____

▶ *For more information about this topic, see pages 204-205 in Pre-GED Science. Answers begin on page 71.*

Solutions and Suspensions: They're Not Always Liquids!

When information is organized in a table, it is easier to make comparisons and contrasts. Columns usually have headings that tell what kind of information is in each column. The rows are read from left to right. They list information about specific entries.

A solution is a type of mixture. The substance in a solution in the larger amount is called the *solvent*. The substance or substances in a solution in the smaller amount are called the *solute*. Think of a glass of juice made from a packaged mix. You add water to the juice mix and stir it in. Most of the mixture is water. Water is the solvent. The juice mix is the solute. We say that the solute *dissolves* in the solvent. It becomes a part of a solution. However, the solvent does not have to be a liquid. It may be a gas or a solid.

The table below lists some common solutions and tells what the solvents and solutes are in each one.

Solution	Solvent	Solute
air (gas)	nitrogen (gas)	oxygen (gas)
a type of dental filling (solid)	silver (solid)	mercury (liquid)
syrup (liquid)	water (liquid)	sugar (solid)
oil paint (liquid)	linseed oil (liquid)	pigments (solid)

■ *Use the passage and table above to answer **Items 1 to 3**.*

1. In a solution, which substance is present in greater quantity? ____________________

2. Name a solution made up of two gases. What are the solvent and solute?

   ____

3. Which solution is made up of a solid solvent and a liquid solute? In this solution, which substance is present in larger amounts?

   ____

► *For more information about this topic, see page 206 in Pre-GED Science.*
Answers begin on page 71.

Unit 3: Optional Writing Activity

■ *Choose one of the writing suggestions below. Write a paragraph on that topic.*

1. Review Lesson 9, Exercise 1, *Atoms and Molecules* on page 40. A subatomic particle, an atom, and a molecule are different from each other. How are they different? Do they have anything about them that is alike? If so, explain what is alike.

2. Picture a small pan of water on a stove. As the water begins to boil, the lid of the pan starts to rattle and move upward. Use what you know about gas laws or review Lesson 10, Exercise 1, *Three States of Matter* on page 44 to explain the movement of the lid.

3. A colander is a bowl with many small holes. It is used in the kitchen. How is a colander similar to the filter paper described in Lesson 11, Exercise 3, *Mixtures* on page 50?

► *Answers begin on page 71.*

Unit 4 **LESSON 12 • EXERCISE 1**

Force and Work: Flying Forces

When you *contrast* things you see how they are different. Some passages contain words that indicate a contrast, but others do not. In the latter case, you will need to pick out key terms and then see how they are different.

Unbalanced forces produce motion. For example, if a forward push on an object is greater than a backward push, the forces are unbalanced. The object will move forward. If the backward push is greater, the object will move backward.

If you study the effects of forces, you will understand how a heavy airplane can fly through the air. The forces applied to an airplane come in pairs. For example, the engines produce a forward force call *thrust*. This is opposed by a backward force called *drag*. Drag is caused by the airplane rubbing against particles in the air.

An airplane's wings produce an upward forward force called *lift*. Lift is produced when air passes at different speeds over and under the wings. This upward force is opposed by *gravity*. Gravity, as you know, pulls objects toward the center of Earth. These two sets of forces, thrust/drag and lift/gravity, help an airplane take off and fly through the air.

■ *Complete* ***Items 1 to 4*** *by filling in the blanks.*

1. Unbalanced forces produce ________________.
2. Of the force pair *thrust* and *drag*, ________________ pushes the airplane through the air and ________________ slows the airplane down.
3. The upward force ________________ is produced when air passes over and under the airplane's wings. However, the force of ________________ pulls the plane toward Earth.
4. An airplane rises into the air because the force of ________________ is greater than the force of gravity.

■ *Use the space provided to answer* ***Item 5.***

5. If the forces of thrust and drag were balanced, what do you think would happen to the airplane? Explain your answer. (Hint: Think of the push and pull forces.)

____

____

► *For more information about this topic, see pages 210–211 in* Pre-GED Science.
Answers begin on page 71.

Force and Work: More or Less Work?

In science, you often need to apply new information to a situation presented. Read the information carefully. What did you already know? What is new information? If there are any formulas, review them. Then apply any new information, plus what you already know, to the situation presented.

> To do work, you have to apply a force to an object. The force you apply must move the object. For example, if you push against the back of a car and it moves forward, you've done work. But if the car doesn't move, no work has been done.
>
> The amount of work you do depends on two things. The first is the amount of force you apply to an object. The second is the distance the object moves. The greater the distance, the greater the amount of work. For example, you do work if you lift a ½-pound book 3 feet. If you lift the same ½-pound book 6 feet, you do twice as much work. If you only lift the book 1 ½ feet, you do half the amount of work.

■ *Use the information above to answer **Items 1 to 5**.*

1. You push a car a distance of 6 feet. Using the same force, your friend pushes the car a distance of 12 feet. Which of you has done more work? ______________

2. You push a car 6 feet. You stop. You try to push again. You push very hard but the car won't move. Which time did you do work? ______________

3. You stand back and look at the car. It begins to roll without your touching it. Did you do any work? ______________

4. You push a car 3 feet. Your friend has to push harder against the car to move it the next 3 feet. Which of you has done more work? ______________

5. Which of these situations involves the most work? Explain.
   (a) You carry a 10-pound bag of flour 20 feet.
   (b) You carry two 10-pound bags of flour 20 feet.
   (c) You carry a 10-pound bag of flour 10 feet.

   ____

▶ *For more information about this topic, see pages 210–211 in Pre-GED Science. Answers begin on page 71.*

Simple Machines: How a First-Class Lever Makes Work Easier

Pictures and diagrams can help you visualize a situation. To understand a picture or diagram, first look at it closely. Get a general idea of what is being shown. Then read any labels. The labels point out important details.

Suppose you needed to lift a 300-pound rock off the ground. You would have to apply a force equal to 300 pounds to the rock. Most people cannot do this. They would need a machine to help them.

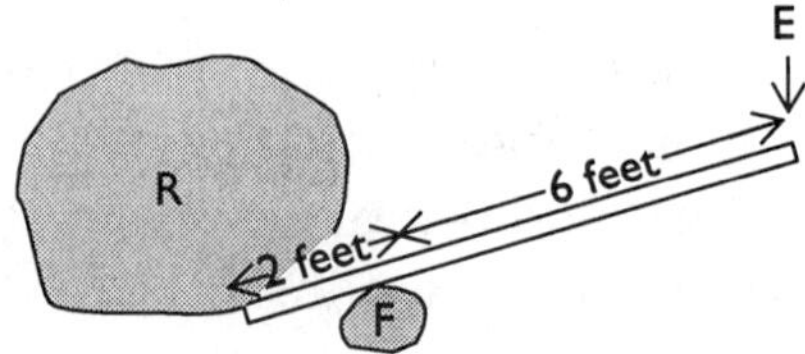

R = resistance force
E = effort force
F = fulcrum

A first-class lever can multiply the force you apply to an object. Look at the drawing. It shows a first-class lever being applied to the rock. The object you want to move, the large rock, is the *resistance force* (R). The force used on the rock is the *effort force* (E). The *fulcrum* (F) is the small rock. A fulcrum is the support, or fixed point, on a lever.

To find out how much the lever increases your effort force, you must divide. You must know the distance from your hands (the effort) to the small rock (fulcrum): in this case, 6 feet. You must also know the distance from the fulcrum to the rock (the resistance): in this case, 2 feet. 6 divided by 2 equals 3. This means that the lever has increased your force by a factor of 3. You would only need 100 pounds of force to move the 300-pound rock.

■ *Use the passage and the illustration above to answer **Items 1 and 2**.*

1. What is the resistance force? What is the effort force?

____

2. Suppose the distance from the effort force to the fulcrum were 8 feet. The distance from the fulcrum to the resistance force remains 2 feet. How many times is the effort force multiplied now? Will it be easier or harder to move the large rock?

____

► *For more information about this topic, see pages 212–213 in Pre-GED Science. Answers begin on page 71.*

LESSON 12 • EXERCISE 4

Momentum: How Much Momentum?

When you make a prediction, you tell what you think will happen. Predictions are usually based on what has happened in the past.

The *momentum* of an object depends on its mass and its speed. For example, think of a car moving slowly down a street. If you drove the same car faster, its momentum would increase. Suppose you see an 18-wheel truck and a small car moving at the same speed. Because its mass is greater, the truck would have more momentum than the compact car.

Strange as it may seem, a football coach has a great interest in momentum. The drawings show three different situations in which a ball carrier and a defensive player meet at a goal line. If the ball carrier crosses the goal line, he will score a touchdown. Note carefully (1) how much each player weighs and (2) how fast each player is moving.

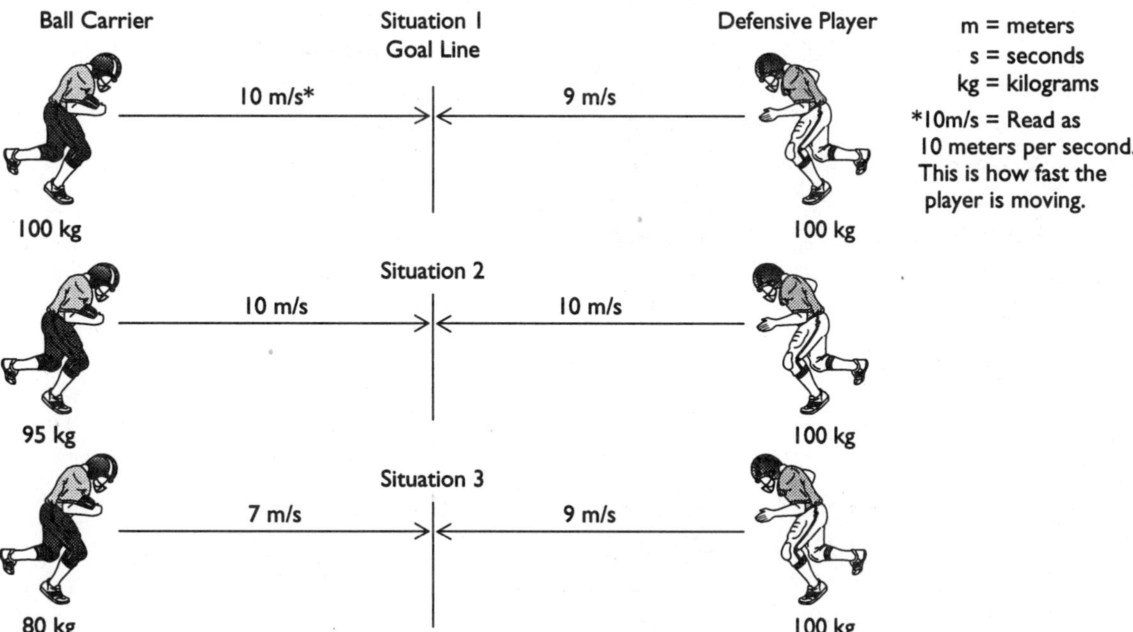

m = meters
s = seconds
kg = kilograms
*10m/s = Read as 10 meters per second. This is how fast the player is moving.

■ *Study each drawing above, then answer* ***Items 1 to 3****: Which player will push back the other player? Explain why.*

1. Situation 1 ______________________________

2. Situation 2 ______________________________

3. Situation 3 ______________________________

► *For more information about this topic, see page 214 in Pre-GED Science. Answers begin on page 72.*

Momentum: Friction

Finding cause and effect is important for understanding some relationships. Remember that causes make things happen. Effects are results. Without the cause, there can be no effect.

Newton's First Law of Motion says that an object at rest remains at rest until an unbalanced force acts on it. Also an object in motion stays in motion until an unbalanced force acts on it. What kinds of forces act on objects to slow them down or stop them? One of the forces is *friction*. Friction is caused as the surfaces of two objects pass across each other.

A dry bar of soap is easy to pick up. But a wet bar of soap can slip out of your hands easily. Why? The water on the wet soap makes its surface very smooth. There is little friction between it and your hand. If you pick the soap up with a dry washcloth, you can hold the soap more easily. The rough surface of the washcloth provides more friction.

The force of friction produces heat. Did you ever rub your hands together on a cold day? The friction between your hands produces heat and your hands feel warmer. Friction also causes wear and tear on objects. The heels on your shoes wear out because of friction. All machines with moving parts become worn due to friction. To reduce the friction and the wear on the parts, the machines are oiled. Less heat is produced and the parts last longer.

■ *Use the information in the passage above to answer **Items 1 to 3**.*

1. What effect does friction have on a moving object?

____

2. The last paragraph tells you two results of friction. What are they?

____

3. Think about walking on a city sidewalk and then on a grassy field. Which will cause greater wear on the heels of your shoes? Why?

____

► *For more information about this topic, see page 214 in Pre-GED Science. Answers begin on page 72.*

LESSON 13 • EXERCISE 1

Energy: From Coal to Toaster Oven

A diagram can make the individual steps of a complicated process very clear. Look carefully at each individual step. Read the labels. Be certain you understand one step before going on to the next step.

Under most conditions, energy cannot be created or destroyed. But it can change form. *Light*, *heat*, *chemical*, and *mechanical energy* can be changed from one form to another. In addition, all forms of energy can be classified in one of two ways. Stored energy is called *potential energy*. The energy of motion is called *kinetic energy*.

This drawing shows a number of steps in a process that brings electricity to a toaster oven. Notice how the energy changes form. Also, notice how it starts out as potential energy and becomes kinetic energy.

Energy Conversions: From Coal to Toaster Oven

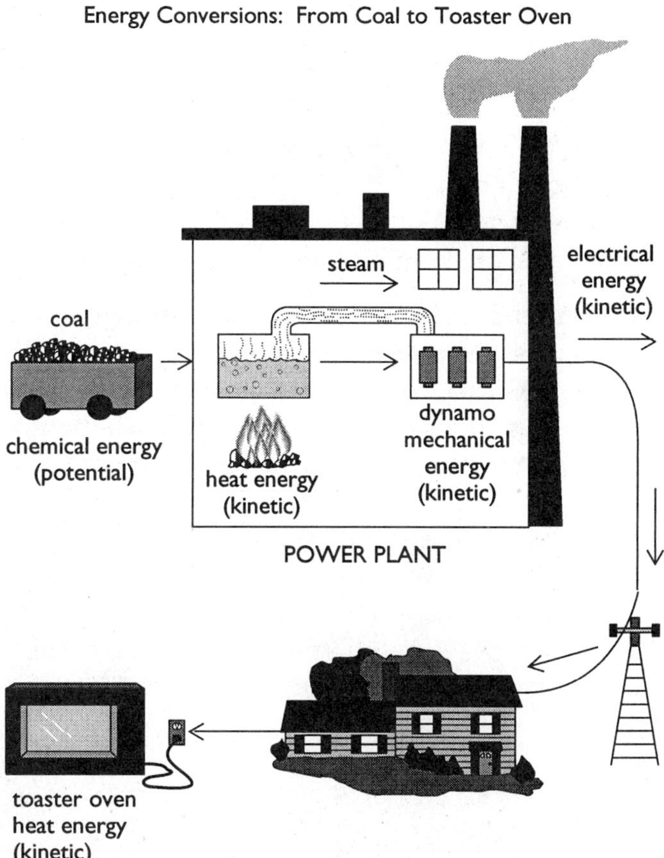

■ *For **Items 1 to 5**, write* T *in the space provided if the statement is true. Write* F *if it is false. If the statement is false, change the underlined word to make the statement true. Write the new word in the space provided.*

_____ 1. The dynamo's movement is an example of <u>potential</u> energy. __________

_____ 2. <u>Chemical</u> energy is stored in coal. __________

_____ 3. When coal is burned, <u>electrical</u> energy is produced. __________

_____ 4. A toaster oven contains <u>potential</u> energy. __________

_____ 5. A power plant converts heat and mechanical energy into <u>electrical</u> energy. __________

► *For more information about this topic, see pages 216–217 in* <u>*Pre-GED Science*</u>. *Answers begin on page 72.*

Energy: Potential Energy, More or Less

Sometimes you are presented with information and are asked to apply that information to a specific situation. Look at the situation carefully. Then apply all you know to answer the questions about the new situation.

One form of potential (stored) energy is produced by *gravity*. Gravity is the downward pull of Earth on an object. If an object falls, it then has the energy of motion, or kinetic energy.

Suppose you are holding a book above the floor. The book would have kinetic energy if you let it go. So there must be energy stored in it. This kind of stored energy is called *gravitational potential energy*. The greater the height of an object, the more potential energy it has.

Something else affects the gravitational potential energy of an object: its mass. Suppose two books are the same distance from the floor. A book with a mass of two kilograms has more potential energy than a book with a mass of one kilogram. So the greater the mass of the object, the greater its potential energy.

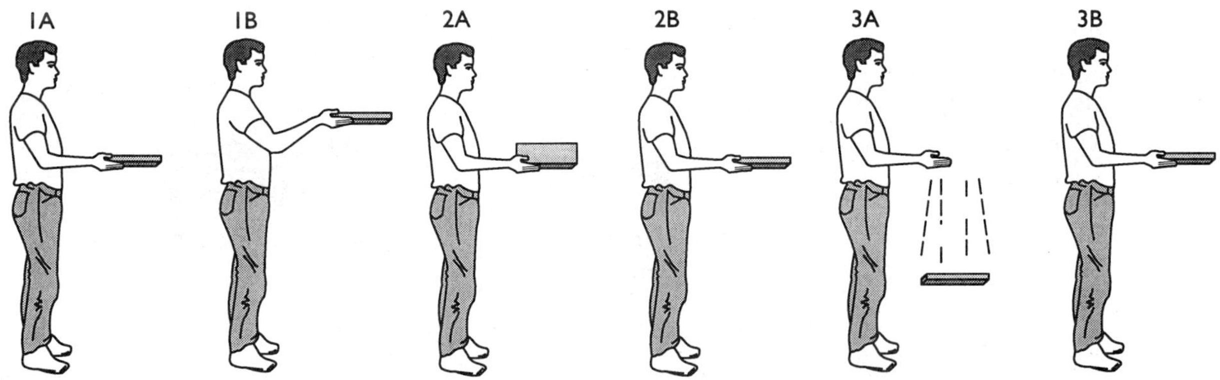

■ *Study the drawings, then answer* ***Items 1 to 3****. (Hint: Think about both height and weight.)*

1. Which book, 1A or 1B, has the greater potential energy? _______________

2. Which book, 2A or 2B, has the greater potential energy? _______________

3. Which book, 3A or 3B, has kinetic energy? _______________

► *For more information about this topic, see pages 216–217 in* Pre-GED Science. *Answers begin on page 72.*

LESSON 13 • EXERCISE 3

Waves: Waves Move Energy, Not Matter

An active reader always takes the time to review. After you finish reading a passage, ask yourself, "What did I learn from the reading?" You may find it helpful to make a list of the important points made or special vocabulary used.

Energy moves from one place to another in waves. A *wave* is a disturbance that moves energy through matter or space. Light and radio waves can move through empty space. That's how sunlight reaches Earth. And it's how radio messages flash from Earth to space probes that are speeding to distant planets.

Other kinds of waves only can travel through matter. For example, sound waves travel through air, water, and many other materials. They cannot travel through empty space. When you watch a ripple on a pond, you may get the idea that the water is moving toward you. However, what actually is moving toward you is not the water. It is energy.

Suppose you toss a pebble into the middle of a pond. What really happens? Some of the kinetic energy of the pebble is transferred to particles of water. This makes the particles move. They bump into nearby particles. This gets the nearby particles to move. They, in turn, bump into other particles, which start to move.

In this way, the energy is passed from one set of particles to another. You see this as ripples, or waves, that spread out across the pond. The particles in the water, however, are not spreading out. They are just moving up and down in the same place.

■ *Use the information from the passage above to answer* ***Items 1 to 5****. Choose the term from the list below that correctly completes each of the sentences.*

empty space	up and down	water
energy	ripples	outward in all directions

1. In water, you see the effects of moving energy as ________________.

2. ________________ moves in waves from one place to another.

3. When you toss a stone into a pond, energy moves ________________.

4. As a wave moves through water, the particles in the water move ________________.

5. Sound waves can travel through ______________ but not through ______________.

► *For more information about this topic, see pages 218–219 in Pre-GED Science.*
Answers begin on page 72.

 LESSON 13 • EXERCISE 4

Electricity: Power

An active reader takes time to preview material before he or she reads it. Skim the passage quickly. Look at any titles, tables, or illustrations to see what information they present. Ask yourself, "What do I already know about this subject?" "What do I expect to find out?"

Power is a scientific term. In the case of electricity, it is a measure of the rate at which electric energy is used. The unit of electric power is called the *watt*, sometimes written W. If you examine any electrical appliance, its power rating is written on it in watts. For example, an electric light bulb may have 60W, 75W, or 100W written on it.

An electric bill is based on the amount of electric energy used each month. Appliances with high power ratings use more electricity than appliances with low power ratings. Therefore, it costs more per minute to run them. It is important to read labels on appliances. That way you will have some idea of how much electricity they use.

POWER RATINGS OF COMMON APPLIANCES

Appliance	Power Rating (in watts)	Appliance	Power Rating (in watts)
hair dryer	1600	refrigerator/freezer	600
13" television set	75	microwave oven	1500
small electric clock	6	clothes dryer	4000
toaster oven	1400	electric range/oven	2500
small coffee maker	650	dishwasher	2400

■ *Use the passage and table above to answer* ***Items 1 to 5***.

1. What is the power rating of a dishwasher? ________________

3. Which appliance is the least expensive to run? ________________

4. Which would cost you more to run, a 13" television set running for 1 hour or an electric clock running for 20 hours? ________________

5. Is it only large appliances that increase you electric bill greatly? Explain.

____

► *For more information about this topic, see page 220 in Pre-GED Science. Answers begin on page 72.*

LESSON 13 • EXERCISE 5

Electricity: Safety

Review questions may ask you to choose between two terms to make a sentence a fact. If you are uncertain about an answer, look back at the passage. Find the section that gives information about the question. Look for facts that support one of the choices.

Electricity can be dangerous because many functions of your body can be disrupted by an electric shock. If electricity enters your body from outside, the regular beating of your heart can be changed. It may even be stopped.

To avoid having accidents involving electricity:

- Never use an electrical appliance if any part of your body is wet or in water.
- Never use an appliance whose wires are exposed or if the protective insulation is broken or worn out.
- Never place an extension cord under a carpet or rug. It may become damaged without your knowing it.
- Plug only one appliance into an outlet. Overloads are produced by connecting too many appliances to one outlet or extension cord. This can cause a short circuit, fires, or shock.
- Get professional help if an electric outlet or switch needs repair. If they are repaired incorrectly, there is a great danger of fire.
- Unplug electrical appliances before doing anything to them—even changing a light bulb. Never place a tool or finger in the socket of an electric appliance that is plugged in.
- Never replace a fuse with anything other than another fuse. If you do so, there is a great danger of fires.
- Never approach a power line that has fallen down outdoors.

■ *Use the information and table above to answer **Items 1 to 5**. Look at the two possible answers in parentheses. Circle the correct answer.*

1. Electrical shocks (can/cannot) cause death.

2. (Several/One) electrical appliance(s) can be safely plugged into one plug.

3. It is (safe/dangerous) to use electrical appliances around water.

4. Home repairs of electrical appliances, outlets, or switches, are (safe/possibly very dangerous).

5. (Always/Never) unplug an electrical appliance before doing anything to it.

► *For more information about this topic, see page 220 in Pre-GED Science. Answers begin on page 72.*

Unit 4: Optional Writing Activity

■ *Choose one of the writing suggestions below. Write a paragraph on that topic.*

1. Review Lesson 12, Exercise 4, *Momentum* on page 57. You have probably seen a movie or TV program where someone is trying to break down a door. Considering what you know about momentum, explain why it helps to run and hit the door rather than just pushing on the door.

____

____

____

____

____

2. Review Lesson 12, Exercise 5, *Momentum* on page 58. Explain how friction is at work in the following situation: You are at the doctor's office. As you wait, you notice that the carpet is worn only in places where people walk frequently, such as by the door.

____

____

____

____

____

3. Review Lesson 13, Exercise 4, *Electricity* on page 62. Describe how you could reduce your electric bill.

____

____

____

____

____

► *Answers begin on page 72.*

ANSWERS AND EXPLANATIONS

Unit 1

LESSON 1

Exercise 1, page 1

1. c 2. a 3. d 4. b

5. Answers will vary. Here is a sample:

   The organelles, or little organs, perform certain jobs within a cell. This allows the cell to live. The organs perform certain jobs within the body. This allows the body to live.

6. The nucleus is the organelle responsible for cell reproduction. It has the "plans" for the cell.

Exercise 2, page 2

1. cell walls and chloroplasts
2. The answer can include any two of the following: nucleus, cell membrane, ribosomes, mitochondria.
3. The cell's wall supports the cell. It also helps protect the cell.
4. The chloroplasts enable the plant to make its own food from the sun's energy.
5. The chloroplast is an organelle. Chlorophyll is a particular kind of molecule within the chloroplast.

Exercise 3, page 3

1. Possible answers mentioned in the passage include red blood cells, muscle cells, and nerve cells. Any normal body cell could be listed, for example: lung cells, brain cells, or skin cells.
2. Cancer cells do not perform specialized jobs.
3. Both grow and reproduce.
4. breathing
5. cigarette smoking, UV rays in sunlight

Exercise 4, page 4

1. white blood cells
2. a virus called HIV
3. They produce antibodies that kill microorganisms.
4. The HIV kills T-cells. When enough disease-fighting T-cells have been killed, the body is no longer able to resist disease.

LESSON 2

Exercise 1, page 5

1. the number of breaths taken per minute; the number of minutes covered by the graph
2. 17 breaths per minute
3. 24 breaths per minute
4. The person was probably walking. Walking takes more energy than sitting but less energy than running. Thus you would breathe more than when sitting but less than when running.

Exercise 2, page 6

1. right atrium, right ventricle, left atrium, left ventricle—list in any order
2. right atrium
3. the lungs
4. keep blood from flowing backward (or to keep blood flowing in one direction only)
5. left ventricle
6. pump blood through the blood vessels

Exercise 3, page 7

1. T
2. F; 120/80
3. F; fall
4. F; high

Exercise 4, page 8

1. shoulder
2. knee
3. ball-and-socket
4. hinge
5. a pivot joint

Exercise 5, page 9

1. contracted
2. by a tendon
3. The muscle bunches up and becomes shorter.
4. A contracted muscle pulls on a bone, causing it to move.

LESSON 3

Exercise 1, page 10

1. the release of an egg from the ovary
2. the shedding of the unfertilized egg and uterine lining through the vagina
3. 28 days
4. the 22nd day
5. the 14th day

Exercise 2, page 11

1. T
2. F; fetus
3. F; the placenta
4. oxygen and food

Exercise 3, page 12

1. dominant
2. recessive
3. two, one from each parent
4. Answer will either be *attached* or *free*.
5. If earlobes are free, the combination may be either dominant-recessive (dr) or dominant-dominant (dd). It cannot be recessive-recessive (rr). If it were, the earlobes would be attached.

Exercise 4, page 13

1. F 2. O 3. F 4. F 5. O

LESSON 4

Exercise 1, page 14

1. meats and dairy products
2. cereals and grains 3. canola
4. It would greatly reduce the amount of saturated fat by about 17%.

Exercise 2, page 15

1. dark green vegetables
2. calcium
3. iron
4. It helps the body store energy.

Exercise 3, page 16

1. the breathing tubes in the lungs
2. In any order: the muscles around the breathing tubes contract and make the tubes narrower, the lining of the tubes becomes inflamed and swollen, making them more narrow, and mucus production may increase.
3. Any three in any order: pollen from plants, dust in the air, mold in the air, tiny hairs from pets, and tiny pieces of skin from pets.
4. In severe asthma attacks, breathing tubes could be completely blocked. If this happens and is not treated immediately, death will result. While this is rare, it is possible.

Exercise 4, page 17

1. A blood vessel may be blocked by a clot or by fatty deposits. A blood vessel may break.
2. F; chest
3. F; panic
4. T

Exercise 5, page 18

1. Jackie, 55 beats/min.; Monica, 75 beats/min.
2. Jackie, 3 minutes; Monica, 4 minutes (Note that jogging starts at 6 minutes, not 0 minutes.)
3. Jackie, 105 beats/min.; Monica, 150 beats/min.
4. Jackie; Jackie's heart did not have to beat as often as Monica's to do the same work. It worked less, or was more efficient.

LESSON 5

Exercise 1, page 19

1. in
2. less
3. muscle
4. outward
5. lungs

Exercise 2, page 20

1. 32°C
2. 16°C
3. The graph would show lower rates over the range of temperatures shown. If there were less sunlight, the rate of photosynthesis would decrease.

Exercise 3, page 21

1. F; cells that contain chlorophyll
2. T
3. F; presence
4. F; oxygen
5. F; glucose

Exercise 4, page 22

1. methane and ammonia
2. The sunlight broke water molecules apart. Water is made of hydrogen and oxygen. The lightweight hydrogen escaped into space. The oxygen was left.
3. the ozone layer
4. simple organisms in the early oceans that performed photosynthesis

Exercise 5, page 23

1. Carbon dioxide levels increase.
2. An increase in artificial sources of carbon dioxide, as, for example, in the burning of fossil fuels.
3. It causes the temperature to increase because less heat is released back into space.
4. Sea levels would rise and coastal lands would flood.

LESSON 6

Exercise 1, page 24

1. a thin shell of land, water, and air where life exists on Earth
2. A community consists of different kinds of living things; a population consists of a single kind of living thing.
3. It contains food and shelter for all the living organisms in it.

Exercise 2, page 25

1. grass
2. mice, frogs, sparrows
3. mice, sparrows

Exercise 3, page 26

1. about 1800
2. about 2000
3. from 1900 to 2000 (twentieth century)

Exercise 4, page 27

1. 1960–1970
2. 1950–1960
3. 1980–1990
4. It will probably be longer. If the increase continues, land changed from rural to urban will probably be higher than in 1980–1990.

Optional Writing Activity, page 28

1. HIV infects white blood cells that protect the body from disease. As these cells die, the immune system does not work well. It becomes deficient. An infected person can become sick from different infections. At this point they have Acquired Immune Deficiency Syndrome, or AIDS.
2. The mother and the fetus are not directly connected. The organ, the placenta, is connected to the mother and the fetus. It is through the placenta that the embryo gets food and air, and gets rid of wastes. Thus you can say that the placenta supports the life of the fetus. Without it, the fetus would die.
3. Answers will vary. Here are two samples.

   People should not worry about overpopulation. The government will find ways to take care of the environment. Science will find ways to produce enough food to support larger populations.

   Everyone should be concerned with overpopulation. We do not know if Earth can feed such a large increase in population. Most people want parks, forest, and woodlands as places of beauty and relaxation. If land has to be used for housing, it will not be available for other purposes.

Unit 2

LESSON 7
Exercise 1, page 29

1. the crust
2. a crack in Earth's crust
3. shifting or breaking of huge slabs of rock along a fault
4. Answers should include any of the following: houses falling down, roads cracking, and bridges collapsing.

Exercise 2, page 30

1. continental slope
2. continental shelf
3. abyssal plain
4. Answers will vary. Here is a sample:

   Although Hawaii does not rise high above sea level, it is at the top of a mountain. The base of the mountain is on the sea floor, deep under the ocean. Thus, one could say that people on the island of Hawaii live on a mountain.

Exercise 3, page 31

1. about 80 gallons
2. drinking and cooking
3. watering the lawn
4. flushing the toilet
5. Answers will vary. It is clear from the graph that the largest use shown, lawn watering, is not essential. One could infer that usage could be cut without harm to human health.

Exercise 4, page 32

1. Everyone uses minerals. They are in many important manufactured products.
2. Tungsten is used in the light bulb and copper is used in the electric wiring.
3. They are all used to make jewelry.

Exercise 5, page 33

1. about 315 parts per million
2. about 1978
3. about 353 parts per million
4. The graph shows that the carbon dioxide level of the air increased. This implies that the temperature also increased.

LESSON 8
Exercise 1, page 34

1. summer
2. winter
3. fall and spring; notice that on March 20 and September 20 there are about 12 hours of daylight, which means that there must be about 12 hours of darkness or night.
4. increases

Exercise 2, page 35

1. The movement of Earth and the moon cause us to not always see the entire illuminated half of the moon.
2. One-half of the moon is lighted, the half facing the sun.
3. the waning crescent phase
4. The new moon stage occurs. This is because none of the illuminated side can be seen. It is facing the sun.

Exercise 3, page 36

1. Saturn
2. Earth
3. Venus, at 243 days
4. Mars, at 24.5 hours

Exercise 4, page 37

1. galaxy
2. star
3. galaxy, universe
4. star
5. universe
6. universe, galaxy
7. galaxy, universe

Exercise 5, page 38

1. blue-white and white
2. red
3. Sirius
4. yellowish
5. color

Optional Writing Activity, page 39

1. Answers will vary. All answers should emphasize avoiding waste. Here are samples:

   Turn off faucets properly so they do not drip; buy water-efficient appliances; run a full load in a washing machine or dishwasher rather than frequent half-empty loads; shower for one minute less each day (saves 6 hours of shower water a year); use water from rinsing fruit and vegetables for watering house plants.

2. Answers will vary. Here is a sample:

   Every day I use toothpaste made with aluminum. I put on my favorite gold jewelry, and I watch TV, which is made using copper and silicon. If I didn't have these items, my teeth would decay, I wouldn't feel as dressed up going out, and I would have to go to a friend's house to watch my favorite TV shows.

3. Answers will vary. Here is a sample:

   People are glad of the chance to get outdoors after a day's work.

Unit 3

LESSON 9
Exercise 1, page 40

1. subatomic particle
2. molecule
3. atom
4. large group of atoms or molecules
5. large group of atoms or molecules

Exercise 2, page 41

1. An atom of one element is unlike an atom of any other element.
2. The atoms of each different element contain a different number of protons.
3. Answers will vary but should include the following idea: Each element looks different.
4. Answers will vary but should include the following ideas: An atom of one element is unlike the atom of any other element. Each element has a set number of protons in its atoms. This difference in atoms causes the elements to be different from each other.

Exercise 3, page 42

1. platinum
2. 907°C
3. Mercury. It melts at -38.9°C.
4. Probably gold would be easiest to identify. You would recognize it by its gold color. If you knew that mercury is a liquid at room temperature, you might choose it. Only mercury would be liquid.

Exercise 4, page 43

1. radon, because atoms become larger as you go down the column
2. chlorine, because it is closer to the top of the column
3. iodine and astatine
4. Elements in groups combine with other elements in similar ways.
5. Chemically stable means that an element does not combine with other elements easily.

LESSON 10
Exercise 1, page 44

1. temperature, pressure
2. space
3. increase
4. decrease
5. Pressure. As you add more and more air, the gases press harder and harder against the sides of the balloon. Finally the pressure is too strong for the balloon. It breaks.
6. In cold weather the volume of a gas decreases. Air takes up less room. So you have to put more air in the tires.

Exercise 2, page 45

1. in Denver, Colorado
2. at about 102°C
3. about 9,000 meters
4. decrease
5. decreases
6. increase
7. higher

Exercise 3, page 46

1. P 2. P 3. C 4. P 5. P
6. C 7. C 8. P 9. P 10. C

Exercise 4, page 47

1. f 2. c 3. a
4. d 5. e 6. b
7. Catalysts speed up chemical reactions. Inhibitors slow down chemical reactions.
8. They speed up vital chemical reactions. Without them, the chemical reactions necessary for life would take too long.

LESSON 11
Exercise 1, page 48

1. two or more
2. soft, bright, silvery
3. greenish-yellow, poisonous
4. white solid made of small crystals
5. element
6. compound
7. unlike
8. linked together

Exercise 2, page 49

1. hydrogen and carbon
2. 1; 4
3. -88.3°C
4. 3
5. As the number of carbon atoms increases, the boiling point increases. Note that all boiling points are less than 0°C. Going from -161.5°C to -42.2°C is an increase in temperature.

Exercise 3, page 50

1. Compounds
2. Mixtures
3. Compounds and mixtures
4. Compounds
5. Mixtures
6. Mixtures

Exercise 4, page 51

1. They are both clear liquids. Beaker 1 contains only water. Beaker 2 contains water with salt in it.
2. Beaker 1 is empty. There is a solid substance at the bottom of Beaker 2.
3. Answers will vary but should include the following information: Beaker 2 contained a mixture. When the water in Beaker 2 evaporated, there was a substance left at the bottom of the beaker. This substance is salt. The mixture has been separated by physical means.

Exercise 5, page 52

1. the solvent
2. air; the solvent is nitrogen; the solute is oxygen.
3. one kind of dental filling; silver, the solvent

Optional Writing Activity, page 53

1. Answers will vary. Here is a sample answer:

   An atom is made of subatomic particles. These particles include protons, neutrons, and electrons. Molecules are made up of two or more atoms linked together. You could say that atoms and molecules are alike in that both are made of protons, neutrons, and electrons.

2. As water begins to boil, it changes to water vapor, a gas. The gas molecules push against the lid of the pan, causing it to move.
3. A colander can be used to separate out any pieces of solid matter larger than the holes in the colander. Filter paper does the same thing. However, the holes in the filter paper are too small to see.

Unit 4

LESSON 12
Exercise 1, page 54

1. motion
2. thrust; drag
3. lift; gravity
4. lift
5. If the plane were on the ground, it could not take off. If the plane were in the air, it would stop moving and crash. Only unbalanced forces produce motion. If the push and pull are equal, there will be no motion.

Exercise 2, page 55

1. Your friend. If there is more movement with the same amount of force, more work is done.
2. Only when the car moved; for work to occur, force must be applied *and* the object must move.
3. No. You applied no force when the car moved.
4. Your friend. Because your friend had to apply more force, more work was done.
5. (b) This situation moves the heaviest load the greatest distance.

Exercise 3, page 56

1. the 300-lb rock; your hands pushing downward
2. 4 times ($8/2 = 4$); easier; first the force was increased 3 times, now it is increased 4 times.

Exercise 4, page 57

1. The ball carrier pushes back the defensive player. The two masses are equal but the ball carrier has greater speed.
2. The defensive player pushes back the ball carrier. His mass is greater while the speeds are equal.
3. The defensive player pushes back the ball carrier. He has both greater mass and speed.

Exercise 5, page 58

1. Friction slows objects down.
2. Friction produces heat and causes wear and tear on objects.
3. Walking on the sidewalk will cause more wear on the soles of your shoes. The sidewalk is rougher than the grass. There is more friction between the sidewalk and your shoes. Thus, there is more wear and tear.

LESSON 13
Exercise 1, page 59

1. F; kinetic
2. T
3. F; heat
4. F; kinetic
5. T

Exercise 2, page 60

1. 1B; the books are the same size, but book B is higher.
2. 2A; the books are at the same height but book 2A weighs more.
3. 3A; since book 3B is not moving, it has no kinetic energy. It only has potential energy.

Exercise 3, page 61

1. ripples
2. Energy
3. outward in all directions
4. up and down
5. water; empty space

Exercise 4, page 62

1. 2400 watts
2. the clothes dryer; It has a higher power rating than any other appliance listed.
3. the small electric clock; It has the lowest power rating.
4. an electric clock; While it uses fewer watts, it is being run 20 times as long.
5. Not necessarily. While very large appliances have the highest power ratings, frequently they are not used for long periods of time. However, using an appliance with a low power rating for a long period of time can add up watts. Leaving lights on uses a lot of electricity, especially if they are 100W or above.

Exercise 5, page 63

1. can
2. One
3. dangerous
4. possibly very dangerous
5. Always

Optional Writing Activity, page 64

1. Momentum depends on both mass and speed. Running increases the speed. If a person can increase momentum enough, he or she will produce a force greater than that of the door. The door will move.
2. Answers will vary but should include that the friction between the carpet and people's shoes causes wear and tear. More people walk in high-traffic areas, wearing out the carpet more quickly.
3. Answers will vary but should include some of the following ideas: Electric bills can be reduced in several ways. Most importantly, electric appliances—and especially lights—should not be left on if they are not being used. Before making new purchases, check the power ratings: There can be big differences in similar products. It may actually cost less in the long run to buy a new, energy-efficient appliance than to keep an old one.

Science Posttest

The Pre-GED science posttest that follows is similar in format to the Skills Preview in the *Pre-GED Science* book. Part I contains items in the same format as many of the exercises you did throughout the book. Part II contains items in the same form as the GED Science Test. In addition, Part II is one-half the length of the actual GED Science Test.

To make the best use of the posttest, follow these steps. First, review the test-taking tips on page 6 of the *Pre-GED Science* book. Then take the posttest. Check your answers using the Answers and Explanations section on pages 85–87. Finally, use the Referral Charts on page 88. The Referral Charts will help you analyze your performance. Completing the Referral Chart is an important part of the test-taking process.

PART I

■ *Read each passage, and answer the questions that follow.*

For ***Items 1 to 6****, place the steps the scientist took in the correct order. Write the numbers* 1 *through* 6 *in the blanks to show the order.*

In 1961, a scientist cloned a frog. A *clone* is an identical living copy of another living thing. To do this, an unfertilized egg cell was taken from an adult, female frog. The nucleus of the egg cell was destroyed. The scientist then took the nucleus from an intestinal cell of a tadpole, or very young frog. Like all body cells, the intestinal cell's nucleus contained all the genetic information needed to make a new frog. This nucleus was placed inside the egg cell.

The egg cell, with its new nucleus, began to divide. A new tadpole began to develop. What was special about this tadpole? It was an identical twin of the tadpole from which the intestinal cell had come. It was a clone of the original tadpole.

_____ 1. The intestinal cell nucleus was placed inside the egg cell.

_____ 2. The scientist destroyed the nucleus of the egg cell.

_____ 3. The new tadpole was a clone of the old one.

_____ 4. The egg cell with its new nucleus began to divide.

_____ 5. The scientist took an unfertilized egg from a female frog.

_____ 6. The scientist took the nucleus from an intestinal cell of a tadpole.

*Complete **Items 7 to 11** using the information from the graph.*

As a person develops from an infant into an adult, certain changes take place in the person's body. Some of the changes are visible, but others are not. The graph below shows how the pulse rate of an average person changes from birth to age 18. Study the graph. Read the labels on the left side and the bottom. Be certain you understand what information the graph presents.

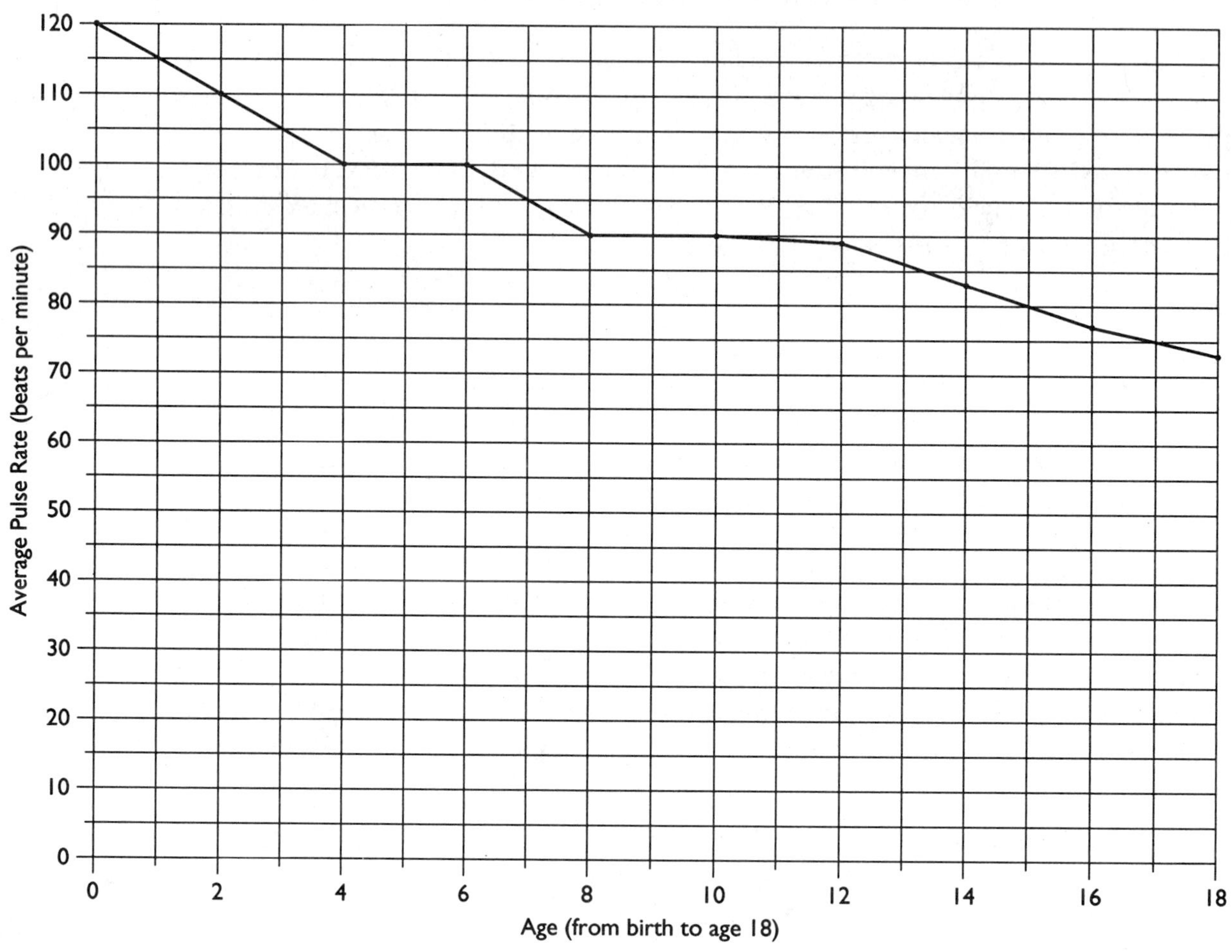

7. The average pulse rate at birth is ________________ beats per minute.

8. The average pulse rate remains at 100 beats per minute between the ages of

________________ and ________________.

9. From birth to 10 years old, the average pulse rate goes down ________________ beats per minute.

10. The average pulse rate of an 18-year-old is about ________________ beats per minute.

11. In general, a person's pulse gets ________________ as the person grows older.

*Answer **Items 12 to 16** using the information and the chart below.*

How was oxygen discovered? Strangely, it involved a series of experiments using a mouse, a candle, and a mint plant. These experiments were performed more than 200 years ago by Joseph Priestley, an English chemist. Here's what Priestley did and observed.

STEPS	WHAT PRIESTLEY DID, OBSERVED, THOUGHT, OR CONCLUDED
1	Priestley placed a burning candle in an airtight container. After a short while, the candle went out.
2	He placed a mouse in another airtight container. Soon the mouse died.
3	Priestley thought that the burning candle and the mouse used up something in the air that they both needed. The candle needed this "something" to keep burning. The mouse needed it to keep living.
4	Priestley was puzzled. He asked himself: Why didn't the animals and all the fires on Earth use up this "something" in air?
5	Priestley thought that somehow nature was replacing what the animals and fires were taking out. He guessed that green plants were doing the replacing. Scientists call such a guess a hypothesis.
6	To test his hypothesis, Priestley put a mint plant into a container where a candle had burned out. No air was allowed to get into the container. Several days later he placed another burning candle in the container. It continued to burn.
7	Priestley concluded that plants put a substance into the air. Fires and animals take out the same substance. Today we know that the substance is oxygen. Oxygen makes up about 21 percent of air.

12. What made Priestley think that a mouse and a candle use up something in air that they both need?

____

13. What puzzled Priestley about what he observed in Steps 3 and 4?

____

14. What was Priestley's hypothesis?

____

15. When Priestley tested his hypothesis, what did he observe?

____

16. What vital substance do green plants add to the air?

____

Answer ***Items 17 to 20*** *using the information and the graph below. Write* **T** *if the statement is true and* **F** *if the statement is false. Correct any false statements by changing the underlined word or words. Write the correct statment in the space provided.*

Without water, life on Earth would not be possible. Some living organisms can live in salt water. Others, like human beings and most land animals, need fresh water. How much of Earth's water is salty? How much is fresh? And how much of Earth's fresh water is easily available? Look at the pie graph to find the answers. It shows what percent of Earth's water is salt or fresh. It also shows where Earth's fresh water is found.

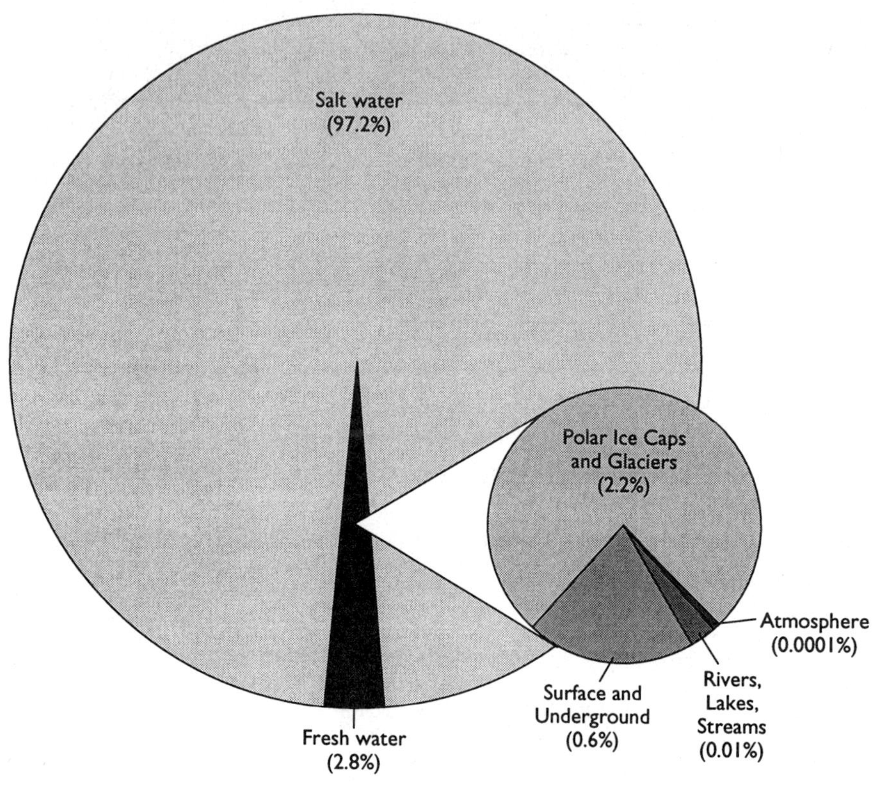

_____ **17.** <u>97.2 percent</u> of Earth's water is fresh.

____

_____ **18.** Most of Earth's fresh water is found in <u>polar ice caps and glaciers</u>.

____

_____ **19.** The smallest amount of Earth's fresh water is found in <u>rivers, lakes, and streams</u>.

____

_____ **20.** Less than <u>1 percent</u> of Earth's water is available as fresh water for human use.

____

PART II

■ *Choose the one best answer to each item.*

***Items 1 to 5** refer to the following information and table.*

Vitamins are found in many foods. You need vitamins to stay healthy. The chart below lists vitamins and some of the foods in which they are found. It also tells why you need the vitamins.

Vitamin	Foods	Use in Body
A	milk, butter, cheeses, fish-liver oils, liver	healthy skin, eyes, bones, teeth
B_1	meat, milk, soybeans, whole grains, nuts, eggs, green vegetables	healthy heart, nervous system, digestive system, muscles
B_2	meat, fowl, milk, nuts, soybeans, eggs, green vegetables, yeast, whole grains	healthy skin, repair of tissues
B_{12}	green vegetables, fish, liver, kidney	healthy blood and nervous system, normal growth
C	citrus fruits (oranges, lemons, limes, grapefruit), leafy green vegetables, tomatoes	normal growth, healthy blood vessels, bones, teeth, healing of wounds, resistance to infection
D	fish-liver oils, milk, eggs, liver	growth, healthy bones and teeth
E	wheat germ, leafy green vegetables, meat, whole grain cereal, milk, butter	healthy cell membranes, healthy reproductive system
K	leafy green vegetables, egg yolk, milk	normal clotting of blood, healthy liver

1. Which of the following is a rich source of vitamin A?
   (1) lemons
   (2) nuts
   (3) soybeans
   (4) cheeses
   (5) meat

2. Which of the following vitamin aids in normal blood clotting?
   (1) A
   (2) B_{12}
   (3) C
   (4) E
   (5) K

3. Limes are a rich source of what vitamin?
   (1) B_1
   (2) B_2
   (3) C
   (4) D
   (5) E

4. If your wounds don't heal, you may not be getting enough of which vitamin?
   (1) B_{12}
   (2) C
   (3) D
   (4) E
   (5) K

5. Vitamin D is needed for which purpose?
   (1) healthy skin
   (2) healthy nervous system
   (3) healthy bones
   (4) healthy liver
   (5) healthy heart

***Items 6 to 10** refer to the following information.*

Rain is turned into acid when certain chemical compounds get into the air. One of these compounds is sulfur dioxide. This compound is made up of the elements sulfur and oxygen. Other chemicals are nitrogen oxides. These chemicals are made up of the elements nitrogen and oxygen. All of these chemicals are gases.

Sulfur dioxide is given off mainly by electric power plants that burn coal. When the sulfur dioxide gets into the air, it changes into sulfuric acid. Nitrogen oxides are given off mainly by cars and other vehicles that burn gasoline. When these oxides get into the air, they change into nitric acid. Sulfur dioxide and nitrogen oxide are also given off by industries that burn coal and oil.

The two chemical compounds make precipitation (rain, snow, and sleet) acidic. Acidic precipitation can kill trees. It can also make lakes so acidic that fish die. Acidic precipitation can also irritate people's eyes, lungs, and skin.

The eastern part of the United States gets more acid rain than other parts. That's because winds generally blow from west to east over the United States. As the winds blow, they carry the gases from power plants, industries, and vehicles in the midwestern part of the country toward the east.

6. Sulfur dioxide is a compound made up of
(1) only sulfur.
(2) only oxygen.
(3) only nitrogen.
(4) sulfur and oxygen.
(5) nitrogen and oxygen.

7. In the air, nitrogen oxides are changed to
(1) oxygen.
(2) sulfur.
(3) sulfuric acid.
(4) nitric acid.
(5) nitrogen.

8. The main sources of sulfur dioxide in the air are
(1) industries that burn coal and oil.
(2) coal-burning power plants.
(3) cars.
(4) rain.
(5) air.

9. What part of the United States gets the most acid precipitation?
(1) far western part
(2) midwestern part
(3) southern part
(4) northern part
(5) eastern part

10. Winds generally blow over the United States from
(1) west to east.
(2) east to west.
(3) north to south.
(4) south to north.
(5) northeast to southwest.

***Items 11 to 15** refer to the following information and illustration.*

Your skeleton is made of 206 bones. While these bones vary in size and shape, they have much in common. All bone tissue is alive. It is constantly being torn down and built back up. The individual bone cells need food and oxygen just like the other cells of your body.

Bone tissue has several functions. Some bones protect internal organs. Others help you move about. Still others support the body. You even have three tiny bones in your ears that allow you to hear.

The skeleton also contains cartilage. This strong connective tissue acts as a cushion in the joints where bones meet each other. Cartilage also is found in the ribs, nose, ears, and between the vertebrae of the vertebral (spinal) column. In the spinal column, the pads of cartilage are usually referred to as disks.

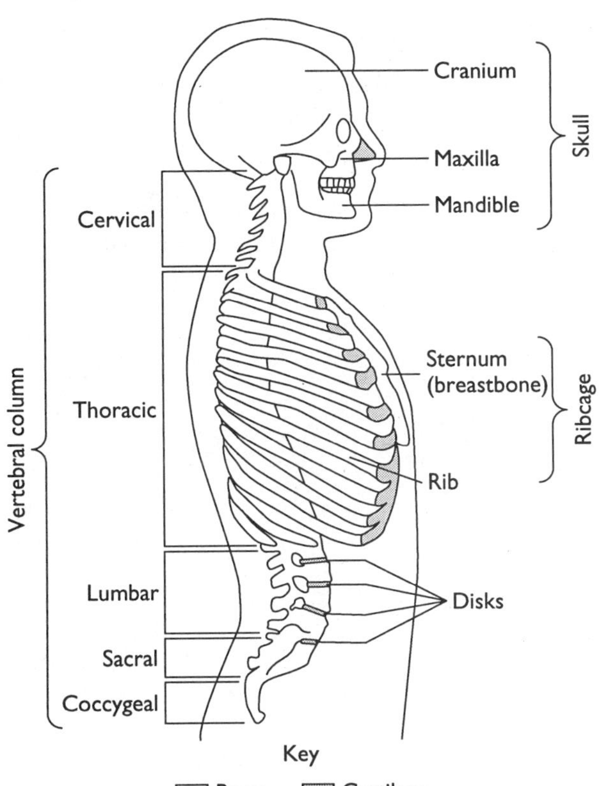

11. Which of the following is *not* true of bone?
(1) It is constantly being torn down.
(2) It is not alive.
(3) It needs both food and oxygen.
(4) It is constantly being built back up again.
(5) It varies in size and shape.

12. Which of the following is *not* mentioned as a function of bone tissue?
(1) Some bone tissue helps you move about.
(2) Some bone tissue helps support the body.
(3) Some bone tissue helps absorb shocks.
(4) Some bone tissue helps protect internal organs.
(5) Some bone tissue helps you hear.

13. All the ribs are attached to the
(1) thoracic vertebra.
(2) disks.
(3) sacral vertebra.
(4) sternum.
(5) lumbar vertebra.

14. The vertebral column has how many main sections?
(1) one
(2) two
(3) three
(4) four
(5) five

15. In the illustration, cartilage is found in the
(1) ribs and vertebral column.
(2) ribs, disks, and nose.
(3) ribs.
(4) cranium, maxilla, and mandible.
(5) only the vertebral column.

***Items 16 to 19** refer to the following information and graph.*

According to a report in *The New York Times,* 40 to 50 million acres of tropical rain forest vanish each year. This is an area about the size of the state of Washington. These forests are cut mostly for timber and to clear land for agriculture. How fast are the tropical rain forests being cut down? The bar graph shows comparative figures for the top five countries during two time periods.

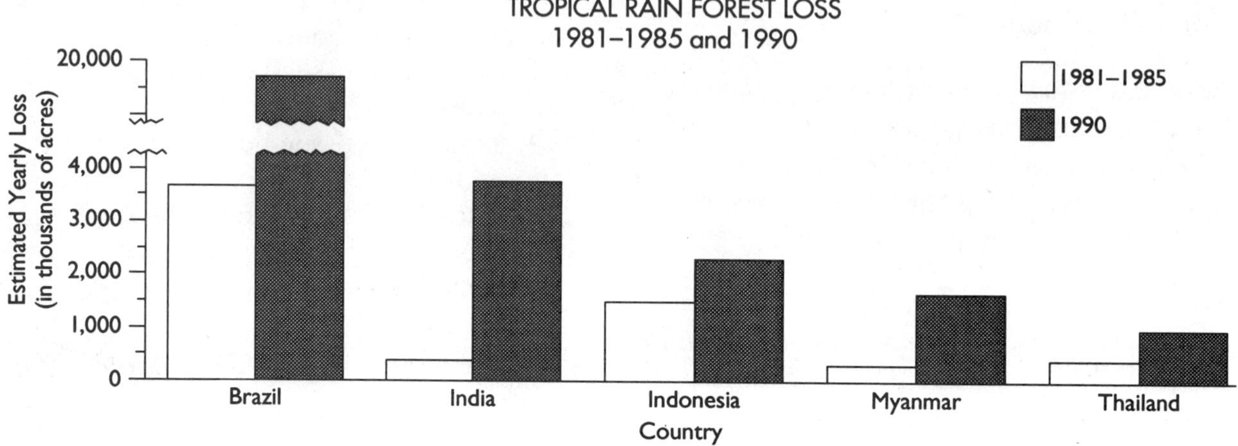

16. Which of the following best describes the information presented in the graph?
    (1) comparison of estimated yearly tropical rain forest loss, 1981–1985 and 1990
    (2) estimated yearly tropical rain forest loss, 1981–1985
    (3) estimated yearly tropical rain forest loss, 1990
    (4) total world tropical rain forest loss from 1981–1990
    (5) world tropical rain forest loss over a period of years, given in thousands of acres

17. Why is the bar for 1990 in Brazil shown broken?
    (1) because the 1990 figures for Brazil are incomplete
    (2) because Brazil lost the largest number of acres of rain forest
    (3) because Brazil's total is so high its bar will not fit in the space allowed
    (4) because Brazil lost the least rain forest
    (5) because it makes the comparison easier to see

18. Which country shows the largest change between their 1981–1985 and their 1990 figures?
    (1) Brazil
    (2) Thailand
    (3) Myanmar
    (4) India
    (5) Indonesia

19. Which country lost almost 4,000,000 (4,000 thousands) acres of rain forest in 1990?
    (1) Thailand
    (2) Brazil
    (3) Myanmar
    (4) India
    (5) Indonesia

***Items 20 to 23** refer to the following information and illustration.*

A pulley is a simple machine made of a rope that fits over a grooved wheel. It makes work easier to do by multiplying (increasing) the force you apply. For example, most people could not lift a 200-pound crate off the ground. But if the right pulley system were attached to the crate, it could be lifted easily.

The number of times a pulley increases your effort is called its mechanical advantage, or M.A. For example, if a pulley increases your effort force 4 times, it has a M.A. of 4. Suppose you pulled with a 50-pound force. The pulley would multiply that force 4 times. Your 50-pound force becomes a 200-pound force.

The drawing shows three pulley systems. Each has a different mechanical advantage. The M.A. of a pulley can be calculated easily by counting the number of lengths of rope that support the object being moved. The number of ropes supporting the object equals the pulley's mechanical advantage. In the illustration, these supporting ropes are marked with arrows.

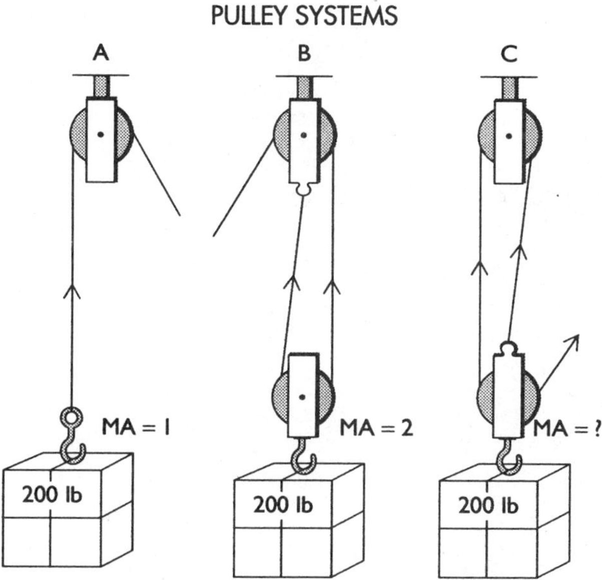

20. Suppose you have a pulley system with a M.A. of 2. You apply a 50-pound force. The pulley system would multiply your force so that it is now
(1) M.A. of 2.
(2) 25 pounds.
(3) 100 pounds.
(4) 200 pounds.
(5) 500 pounds.

21. In a pulley system, the M.A. is determined by
(1) how hard you pull.
(2) the size of the load you want to move.
(3) the number of pulleys in the system.
(4) how many ropes support the load being moved.
(5) your effort force.

22. Which pulley or pulleys will not increase your effort force?
(1) pulley A
(2) pulley B
(3) pulley C
(4) pulley A and pulley B
(5) pulley A and pulley C

23. What is the mechanical advantage of pulley C?
(1) 1
(2) 2
(3) 3
(4) 4
(5) Not enough information is given.

Items 24 to 26 *refer to the following information.*

Soil is made up of weathered rock, decaying plant and animal matter, air, and water. Over the years, the rocks in the soil are broken down. The pieces become smaller and smaller. This process is called *weathering*. These tiny pieces become mixed in with the other parts of the soil.

As plants and animals die, they break down into small pieces. These tiny pieces of decaying matter are called *humus*. Humus has a high carbon content, so it is dark in color. The smell of rich soil is due to the carbon in its humus.

Soil takes hundreds of years to form. All soil is not all alike. In some places, it is only a few inches thick. In other places, it may be several feet thick. The kind of weathered rock in the soil affects its color. It also affects its mineral content. The amount of humus in soil varies greatly. The higher the humus content, the better it is for farming.

24. Soil is made up of
(1) weathered rock and humus.
(2) mostly humus.
(3) mostly weathered rock.
(4) weathered rock, decaying plants, and air.
(5) weathered rock, humus, air, and water.

25. In general, soils are
(1) easily replaced.
(2) formed in a short period of time.
(3) not all alike.
(4) rich in humus.
(5) all alike.

26. The formation of soil
(1) is caused by weathering of rock.
(2) is caused by the decay of plants and animals.
(3) is slowed down by farming.
(4) is a lengthy process.
(5) depends on the kind of rock that weathers.

Items 27 to 29 *refer to the following information.*

Lightning is a type of static electricity. If huge amounts of static electricity build up in thunderclouds, it is discharged in a bolt of lightning. A lightning bolt is made of negatively charged electrons. Some lightning bolts go from cloud to cloud. Other lightning bolts go from clouds to Earth.

When lightning hits Earth, it can be very destructive. Sometimes lightning starts fires. In cities, lightning may hit telephone or power lines. When this happens, the phone service or power may be out for a period of time. Lightning also is dangerous to human life.

Lightning is attracted to the tallest object in an area. Because of this, some people install lightning rods on buildings. In rural areas they can often be seen on barns. A lightning rod is a long, pointed, metal rod. It is placed so that it sticks into the air above a building. A cable connects the rod to the ground. Lightning is attracted to the rod. Its electrons travel along the cable into the ground. Here they are harmlessly spread through the ground. Without the lightning rod, the lightning would hit the building, and most likely cause a fire.

27. A lightning bolt is
(1) made of electrons.
(2) made of negatively charged electrons.
(3) a static discharge from a cloud to the ground.
(4) current electricity.
(5) made of positively charged electrons.

28. Which of the following is *not* mentioned as an effect of lightning strikes?
(1) fires
(2) severe thunderstorms
(3) loss of power
(4) loss of life
(5) loss of telephone service

29. Which of the following is *not* a cause-and-effect relationship?
(1) buildup of static electricity in clouds, lighting bolt discharged
(2) lightning hits power line, loss of electric power
(3) lightning hits lightning rod, lightning spread harmlessly through ground
(4) lightning rods installed, building protected from fire due to lightning strike
(5) lightning is static electricity, lightning bolt made of electrons

***Items 30 to 31** refer to the following information.*

Chemical formulas show what elements a substance contains. They also show how many atoms of a particular element are present. For example, look at the formula CO_2. C is the atomic symbol for carbon, and O is the symbol for oxygen. When we see the formula CO_2, we can tell that it contains one carbon atom and 2 oxygen atoms.

Here are the symbols for three other elements. Cl = chlorine, Na = sodium, and Si = silicon.

30. The chemical formula, $NaClO_3$, contains how many elements?

(1) 3
(2) 4
(3) 5
(4) 6
(5) 7

31. Which of the following formulas does *not* contain the element carbon?

(1) CCl_4
(2) $C_6H_{12}O_6$
(3) $Na2CO_3$
(4) SiO_2
(5) CO_2

***Items 32 to 33** refer to the following information.*

Fossils are remains and traces of once-living animals and plants. Some fossils are the remains of the organisms. Other fossils are traces of them, such as footprints. Fossils can be classified as molds or casts.

Suppose a clam dies and is buried in the mud. Over a period of time the mud hardens into rock. The clamshell dissolves, but it leaves a space, or hollow, in the rock. This space is exactly the same size and shape as the clamshell. It even shows the tiny groves on the shell. A mold has been formed. A mold looks like an imprint of the organism, not like the organism itself.

Now, suppose this mold was filled in by some substance. The substance later hardened. Now you would have a cast. A cast looks like the organism itself.

32. The most important difference between molds and casts is that

(1) molds show remains, but casts show traces.
(2) molds are found in mud, but casts are not.
(3) molds are hollow impressions, but casts look like the original organisms.
(4) molds are animal remains, and casts are plant remains.
(5) molds are not hardened, but casts are.

33. Which of the following would be classified as a plant or animal trace, not a remain?

(1) a fossilized leg bone
(2) a fossilized bird's wing
(3) a rock with fossilized shells in it
(4) a sheet of rock with a fossilized fern leaf
(5) a set of fossilized dinosaur tracks

POSTTEST ANSWERS AND EXPLANATIONS

PART I, Pages 73–76

1. 4
2. 2
3. 6
4. 5
5. 1
6. 3
7. 120
8. 4; 6
9. 30
10. 72.5 exactly; 72 or 73 is acceptable
11. slower
12. After being in a closed container for a while, the mouse died and the candle went out.
13. Priestley wondered why animals and fires didn't use up all of the "something" in the air.
14. Priestley guessed that plants put back into the air what it was that animals and fires took out.
15. Priestley observed that a mint plant seemed to add the substance to the air because the second burning candle did not go out.
16. oxygen
17. F; 2.8%
18. T
19. F; the atmosphere
20. T (Only surface and underground water (0.6%) and rivers, lakes, and streams (0.01%) are available. The rest of the fresh water is frozen or in the air.)

PART II, Pages 77–84

1. (4) (Comprehension) The table shows that cheese is one of the sources of vitamin A. None of the other choices is a source of vitamin A.
2. (5) (Comprehension) The table shows only vitamin K with this effect. All other choices have different uses in the body.
3. (3) (Comprehension) Limes are only listed as a rich source of vitamin C. They are a citrus fruit. Limes are not a rich source of any other vitamins, so the remaining choices are incorrect.
4. (2) (Comprehension) The table shows only vitamin C having this use in the body. All the remaining choices have other uses but not healing wounds.
5. (3) (Comprehension) The chart shows only vitamin D having this use in the body. All the remaining choices have other uses but not for healing wounds.
6. (4) (Comprehension) The first paragraph states that sulfur dioxide is made up of sulfur and oxygen. Choices (1) and (2) are only half of what is necessary. Nitrogen is part of nitric acid, so Choices (3) and (5) are incorrect.
7. (4) (Comprehension) The second paragraph states that nitrogen oxide changes to nitric acid. All other choices are incorrect.
8. (2) (Comprehension) The passage says that sulfur dioxide is given off mainly by electrical power plants that burn coal. Although Choice (1) is a source, it is not the main source. Choice (3) is a source of nitrogen oxide, not sulfur dioxide. Choices (4) and (5) are not sources.
9. (5) (Comprehension) This information is stated in the last paragraph. All other choices do not receive as much acid precipitation.

10. (1) (Comprehension) This information is stated in the last paragraph. The winds do not generally blow in any other direction, so the remaining choices are incorrect.

11. (2) (Comprehension) The first paragraph states that bone tissue is alive. All other choices are true of bone tissue.

12. (3) (Comprehension) Cartilage, not bone, acts as a shock absorber. All other choices are mentioned as functions of bone tissue.

13. (1) (Comprehension) The illustration shows that all of the ribs are attached to the thoracic vertebra. Only some of the ribs are attached to the sternum. Because of this, Choice (1) is better than Choice (4). All remaining choices are incorrect according to the diagram.

14. (5) (Comprehension) The illustration shows five sections: cervical, thoracic, lumbar, sacral, and coccygeal. The other choices are not the right number.

15. (2) (Comprehension) Choice (1) omits the nose. Choice (3) omits the nose and the disks in the vertebral column. Items in Choice (4) all refer to the head, omitting the other locations. Choice (5) omits the ribs and nose.

16. (1) (Comprehension) This is the best description of the graph's information. None of the other choices contain incorrect information. However, they cover only a part of what is on the graph. The most important omission is the fact that the graph shows a comparison.

17. (3) (Comprehension) If the bar for Brazil were shown completely, it would have to be about five times longer than the bar of India in 1990. None of the other choices is correct.

18. (4) (Analysis) The figure for India shows an extremely large increase. It is an increase of approximately eight times. None of the other choices shows such a large increase.

19. (4) (Analysis) India shows a loss of 3,707 thousand acres. None of the other choices shows a loss of almost 4,000,000 acres.

20. (3) (Application) Your 50-pound force multiplied by 2 becomes a 100-pound force. Choice (2) is a result of dividing your force by 2. The remaining choices are incorrect.

21. (4) (Comprehension) The last paragraph says that the number of ropes equals the pulley's mechanical advantage. None of the other choices determines the mechanical advantage (M.A.).

22. (1) (Comprehension) The M.A. of pulley A is 1. If your force is 60 pounds and you multiply it by 1, you still have 60 pounds. The remaining choices are incorrect.

23. (3) (Comprehension) The pulley shows three ropes supporting the object being moved. This gives it a M.A. of 3. The remaining choices do not show the correct M.A. of a pulley system with three ropes.

24. (5) (Comprehension) The information is given in the first sentence of the reading, and the second paragraph states that humus is plant and animal matter. The other choices do not provide the complete ingredients of soil.

25. (3) (Comprehension) This information is given in the last paragraph. Choices (1) and (2) are incorrect. Choices (4) and (5) are not supported by information in the passage.

26. (4) (Comprehension) This information is given in the last paragraph. Choices (1) and (2) are not the only processes that form soil. Choice (3) is not stated or implied in the passage. Choice (5) does not cause formation, it helps determine soil type.

27. (2) (Comprehension) Choice (1) is not wrong but is not the best answer to the question. Choice (3) is true but does not include the fact that lightning can go from cloud to cloud. Choices (4) and (5) are not what make up a lightning bolt.

28. (2) (Comprehension) Although thunderstorms may accompany a storm, they are not mentioned in the passage as an effect of lightning. All other choices may result from lightning strikes.

29. (5) (Analysis) Both parts of Choice (5) are true. However, the fact that lighting is static electricity does not cause lightning bolts to be made of electrons. All other choices are cause-and-effect relationships.

30. (1) (Comprehension) This formula contains the elements: sodium, Na: chlorine, Cl; and oxygen, O. Choice (3) is the number of atoms. All remaining choices are incorrect.

31. (4) (Analysis) All the other formulas contain C, carbon. SiO_2 does not.

32. (3) (Analysis) Only this choice is supported by information in the passage. The other choices are not important differences.

33. (5) (Analysis) Choices (1), (2), (3), and (4) are all actual parts of animals, their remains. Choice (5) indicates that an animal was there, but does not show the animal itself.

SCIENCE POSTTEST REFERRAL CHARTS

After you have completed the Posttest, check your answers against the Answers and Explanations on pages 85-87. On the charts below, circle the items you answered correctly. Then fill in the total number of items you answered correctly.

When you have completed the charts, turn to page 89 to record your results and to assess your readiness to move on to the South-Western *GED Science* book.

Section 2: Lessons	Part II Items	Number of Items Correct
Skill 1: Thinking Like a Scientist		
Skill 2: Becoming an Active Reader	8, 9, 10, 12, 25, 26, 27	
Skill 3: Working with Words		
Skill 4: Finding the Main Idea	16	
Skill 5: Finding Details	6, 7, 24	
Skill 6: Restating and Summarizing		
Skill 7: Reading Tables and Graphs	1, 2, 3, 17, 18, 19	
Skill 8: Reading Diagrams and Maps	13, 14, 15	
Skill 9: Making Inferences	23	
Skill 10: Making Predictions		
Skill 11: Applying Information	11, 20, 21, 30, 31	
Skill 12: Identifying Cause and Effect	4, 28, 29	
Skill 13: Comparing and Contrasting		
Skill 14: Classifying Information	5, 33	
Skill 15: Recognizing Fact, Opinion, and Hypothesis		
Skill 16: Evaluating Information and Drawing Conclusions	22, 32	

TOTAL: ____ of 33

Section 3: Units	Part I Items	Number of Items Correct	Part II Items	Number of Items Correct
Unit 1: Life Science (pp. 136–173)	1, 2, 3, 4, 5, 6, 7, 8, 9, 10, 11		1, 2, 3, 4, 5, 11, 12, 13, 14, 15	
Unit 2: Earth Science (pp. 174–187)	17, 18, 19, 20		16, 17, 18, 19, 24, 25, 26, 32, 22	
Unit 3: Chemistry (pp. 188–207)	12, 13, 14, 15, 16		6, 7, 8, 9, 10, 30, 31	
Unit 4: Physics (pp. 208–221)			20, 21, 22, 23, 27, 28, 29	

TOTAL: ____ of 20 **TOTAL:** ____ of 33

PRE-GED SCIENCE EVALUATION CHARTS

Are you ready for more advanced GED preparation materials? The evaluation charts that follow on page 90 can help you determine your strengths. They can also help identify areas where your skills need further development. Here's how to use them.

1. On page 90, record your scores from each Referral Chart in South-Western's *Pre-GED Science* book and this exercise book. (Page references are given for each chart.)

2. Calculate your total score for each row (across).

3. Total each column (down) to determine your overall performance.

4. Evaluate your performance in each category with your instructor.

   - Use the assessment information below to evaluate your achievement level.

   - Record the date when mastery was achieved for each particular skill or content area.

   - Determine how you should proceed to more advanced GED preparation materials based on these results.

ACHIEVEMENT LEVELS	
Mastery	Your scores are consistently high in this category. You answer these questions with confidence and do not require assistance.
Satisfactory	Your scores in this category are reasonably high but are not consistent. You often must reread questions several times or ask for some assistance.
Fair	Your scores in this category need improvement. Further study and practice are needed.

Completing this evaluation will help you and your instructor determine if and in what areas you are ready to move on to more advanced GED preparation materials. Good luck with your studies as you prepare to pass the GED Tests!

Pre-GED Science Evaluation Charts Name ______________________ Starting Date __________

Section 2: Foundation Skills	Book: Foundation Skills Review (p. 133)	Book: Posttest (p. 246)	Exercise Book: Posttest [Part II items] (p. 88)	Total Score	Date Mastery Achieved
	Record Number of Items Correct				
Skill 1: Thinking Like a Scientist (pp. 22–27)				___ of 3	
Skill 2: Becoming an Active Reader (pp. 28–33)				___ of 7	
Skill 3: Working with Words (pp. 34–39)				___ of 2	
Skill 4: Finding the Main Idea (pp. 40–45)				___ of 5	
Skill 5: Finding Details (pp. 46–51)				___ of 7	
Skill 6: Restating and Summarizing(pp. 52–57)				___ of 3	
Skill 7: Reading Tables and Graphs (pp. 58–65)				___ of 8	
Skill 8: Reading Diagrams and Maps (pp. 66–71)				___ of 6	
Skill 9: Making Inferences (pp. 72–77)				___ of 6	
Skill 10: Making Predictions (pp. 78–83)				___ of 5	
Skill 11: Applying Information (pp. 84–89)				___ of 9	
Skill 12: Identifying Cause and Effect (pp.90–95)				___ of 7	
Skill 13: Comparing and Contrasting (pp. 96–101)				___ of 4	
Skill 14: Classifying Information (pp. 102–107)				___ of 5	
Skill 15: Recognizing Fact, Opinion, and Hypothesis (pp. 108–113)				___ of 3	
Skill 16: Evaluating Information and Drawing Conclusions (pp. 114–119)				___ of 5	
TOTALS:	___ of 19	___ 33	___ of 33	___ of 85	

Section 3: Science Units	Book: Science Review (p. 236)	Book: Posttest (p. 247)	Exercise Book: Posttest [Part II items] (p. 88)	Total Score	Date Mastery Achieved
	Record Number of Items Correct				
Unit 1: Life Science (pp. 136-173)				___ of 32	
Unit 2: Earth Science (pp. 174-187)				___ of 21	
Unit 3: Chemistry (pp. 188-207)				___ of 19	
Unit 4: Physics (pp. 208-221)				___ of 18	
TOTALS:	___ of 24	___ of 33	___ of 33	___ of 90	